D0116143

Collectible Bells

Treasures of Sight and Sound

Donna S. Baker

4880 Lower Valley Road, Atglen, PA 19310 USA

HOUSTON PUBLIC LIBRARY
R01225 54698

Dedication

To Eliza, the four-legged and dearly beloved "belle" of the Baker household.

Library of Congress Cataloging-in-Publication Data

Baker, Donna S.
Collectible bells: treasures of sight and sound/Donna S. Baker.
p. cm.
Includes bibliographical references and index.
ISBN 0-7643-0555-7
1. Bells--Collectors and collecting--United States--Catalogs.
I. Title.
NK3653.B35 1998
681'.868848'075--dc21 98-4707
CIP

Copyright © 1998 by Schiffer Publishing Ltd.

All rights reserved. No part of this work may be reproduced or used in any form or by any means—graphic, electronic, or mechanical, including photocopying or information storage and retrieval systems—without written permission from the copyright holder.

Designed by "Sue"
Typeset in *President/DellaRobbia BT/Futura Bk BT*

ISBN: 0-7643-0555-7
Printed in China
1 2 3 4

Published by Schiffer Publishing Ltd.
4880 Lower Valley Road
Atglen, PA 19310
Phone: (610) 593-1777; Fax: (610) 593-2002
E-mail: Schifferbk@aol.com
Please write for a free catalog.
This book may be purchased from the publisher.
Please include $3.95 for shipping.

In Europe, Schiffer books are distributed by
Bushwood Books
6 Marksbury Avenue
Kew Gardens
Surrey TW9 4JF England
Phone: 44 (0)181 392-8585; Fax: 44 (0)181 392-9876
E-mail: Bushwd@aol.com

Please try your bookstore first.

We are interested in hearing from authors
with book ideas on related subjects.

Contents

Acknowledgments

This book would not have been possible without the assistance and support of many fine people. First and foremost, I would like to thank the collectors who so graciously and generously allowed me to photograph their bells and who shared with me concurrently their knowledge and their enthusiasm for bells: Iva Mae Long and Peggy Long, Larry and Marjorie Glassco, Arlene B. Green, Mr. and Mrs. Byron G. Ward, and Jean Cline. Your contributions to this book were most invaluable and I deeply appreciate your participation.

I was fortunate to meet or talk with numerous members of the American Bell Association during the course of writing this book, and I can attest to the fact that each one lived up to the association's reputation for being friendly and helpful. In particular, I would like to thank Ron and Kay Weaver, Neal Goeppinger, Barbara Smith, Sue and Denny Moore, Eleanor Evans, and Kathryn and Herb Stafford for the wealth of information, advice, and support they provided to me along the way.

Thanks also to my colleagues Jeffrey Snyder and Bruce Waters for their assistance with the photography, data recording, (figurative) hand-holding, and lugging of bulky equipment.

Finally, I would like to thank Peter and Nancy Schiffer and Douglas Congdon-Martin for their support, guidance, and confidence in my abilities throughout the preparation of this book.

Introduction

Dictionary definitions notwithstanding, the task of precisely and succinctly describing "bells" is a daunting one. To specify their dimensions, one needs measurements ranging from inches to yards, ounces to tons. A list of materials used in their manufacture must include metal, glass, clay, wood, and even "recycled" materials from other objects. A roster of how bells are used would be even longer.

Yet perhaps it is this very diversity, this defiance of simple enumeration, that leads to our fascination with bells. The primitive animal bell fashioned from crudely bent metal is as much a bell as the elegant, hand painted porcelain figurine with carefully defined features. And each has in common the singular characteristic that sets bells apart from many other collectibles: we appreciate them for both their visual and aural qualities, that is, they ring!

The history of these complex objects is one that dates back to antiquity. The ancient Chinese were among the first to make and use bells, but their existence has been documented as well in the early cultures of Egypt, India, Greece, and Rome. Once bellfounding was developed in Europe, around the sixth to eighth centuries, metal bells advanced from those basically hammered into shape to those cast in molds that produced more standard shapes and pleasing tones. Bellfounding involves the pouring of molten metal into large molds and allowing it to cool and harden in the mold. The first bellfounders in Europe were monks; occasionally they cast large, heavy bells right in the churchyard, thus avoiding the problem of how to transport them elsewhere when completed. By around the tenth century, bell towers housing massive church bells were commonplace throughout Europe.

In the Middle Ages, bells were an integral part of daily life in towns and cities. From the large booming town bells to the small but stalwart town crier bells, bells let the townsfolk know what was going on. They announced births and deaths, proclaimed the opening of local markets, warned of unexpected danger, and signaled curfew time when residents were required to turn in for the night. Later, bells served to call schoolchildren to class, farmhands to dinner, servants to dining rooms, hotel clerks to desks.

Modern electronics, technology, and contemporary customs have rendered many former bell uses obsolete. We mail out printed birth announcements, schedule our day with clocks and watches, push buzzers outside of doors, carry pagers to be immediately accessible, and monitor mealtime by the ding of the microwave. No matter—this only serves to enhance the attraction of old and timeworn bells that harbor a story or two about their former occupations.

Acknowledging the vast array of collectible bells available today, this book is designed as a broad based survey of their most characteristic forms and types. The medley is an eclectic one, including bells both old and new, invaluable to affordable. Bells of all sizes are shown, however the book focuses primarily on small to medium sized bells that are easily displayed inside a home. While every effort has been made to include bells of interest and appeal to a wide audience, the exclusion of any type of bell or any individual bell is not intended as a reflection of its importance or relative value.

Categorizing bells presents an even greater challenge than describing them, as it is the rare bell that falls neatly into just one category. Following an illustrated review of the materials most commonly used for bells and a mini-tour of bells around the world, this book divides the majority of bells remaining into those that perform an actual function (both appearance and sound are important) and those that are primarily for pleasure (appearance is most important). One could certainly (and successfully) argue that bringing pleasure and enjoyment to the beholder is a legitimate function; after all, our lives would be most drab and barren without beautiful objects to surround and gladden us. Nonetheless, the distinction between bells appreciated most for their specific purpose and bells appreciated most for their pleasing countenance is the one used here, with acknowledgment that other classification systems are equally valid.

Due to their "cross-over" qualities, some broad categories of bells will be found in more than one section. Chinese bells, for example, are shown in Chapter Two as examples of enameled bells, in Chapter Three as representative Asian bells, and in Chapter Four as an illustration of religious bells.

A final word about sound, that essential and fundamental part of any bell. Like the smooth contours of a sculpture reaching out for tactile exploration, bells call out to be rung. The sounds they produce are as heterogeneous as the bells themselves, usually melodic and extraordinarily pleasing, sometimes disappointing or slightly tuneless. Yet the true nature of a bell is fully appreciated only when it is heard as well as seen, and for that you will need to seek out three-dimensional bells. I encourage you to do so, for if you have not already been captivated by the sight and sound of these delightful treasures you will surely find them a most absorbing, fascinating, and rewarding pursuit!

About the Values

The values listed for the bells in this book are intended to provide readers with a general idea of what they might expect to pay for the same or similar bell in today's market. The values represent a guideline only and are not meant to "set" prices in any way. It is entirely possible to purchase a bell for a higher or lower amount than the value shown here, as many factors affect the actual price paid. These factors include the bell's age, condition, workmanship, size, and scarcity. None of these are absolutes, however. Older or larger bells are frequently among the most expensive, for example, yet some contemporary bells of relatively small size may command high prices due to their exceptional quality and limited availability. In addition, geographic location, the context in which the bell is sold, and the buyer's relative desire to own a particular bell can significantly impact final purchase prices.

Chapter One

Bell Basics

What image comes to mind when you hear the word "bell?" Do you picture the traditionally shaped bell with an inverted cup-like form? Or do you think of the small round "jingle bells" that adorn the backs of horses? Both are correct, yet these are far from the only kinds of objects that qualify as bells.

As you will see throughout the pages of this book, bells are amazingly diverse. Although they share many common qualities—most notably the ability to produce sound—bells differ vastly in terms of their size, their shape, what they are made out of, and how the sound is produced. Their name comes from the old Anglo-Saxon word *bellan*, which means "to bellow." The Italian word for bell, *campana*, is the source of the word "campanology," the study of bells, and of "campanile," a bell tower.

Familiar looking cup shaped bells, usually referred to as **open mouth**, range in size from huge church bells weighing thousands of pounds and bearing deep tones that reverberate for many minutes to dainty tea bells and those even smaller that ring with sweet, melodious notes. Open mouth bells most typically produce sound by means of a clapper, a round or oblong object suspended inside by a rod or chain. Ringing occurs when the bell is swung back and forth, causing the clapper to strike against the bell's inner rim. With some large bells, the bell itself remains stationary and the clapper is activated by pulling on it with a rope. Still other open mouth bells, such as large Buddhist bells, are rung by striking the outside of the bell with a rod or hammer.

Large, heavy church bell with characteristic open mouth shape. 48" diameter. $3000-3800.

Open mouth aluminum bell, mounted outdoors. 10.5" diameter. $70-90.

Traditionally shaped Royal Winton ceramic bell, made in Grimwades, England. 4" high. $5-8.

Tall ceramic open mouth bell from Jerusalem, hand painted. 7.25" high. $35-50.

Open mouth paneled glass bell with octagonal shape. 5.25" high. $35-50.

Small tea bell with open mouth shape, bell tower pictured on handle. 3.75" high. $5-8.

Sleigh bells and other closed mouth bells are more appropriately known as **crotals**. Similar to open mouth bells, crotals vary considerably in size and shape, although most are spherical in appearance. They may have a long slit or opening across the bottom or they may be entirely enclosed. One or more pellets inside creates sound when the bell is swung or shaken. In fact, the word crotal comes from the Greek word *krotalon*, meaning "rattle." Some of the earliest bells known to man are of this variety; indeed bronze crotals unearthed from ancient graves in the Luristan area of Iran (formerly Persia) are believed to be horse bells that were used by warriors from the ninth to eighth century BC.

Set of fifty matched sleigh bells riveted to a 6' 10" long leather strap. These crotals are 1.25" stamped steel with the bottom halves clinched over the top halves. $75-100.

Very early "horse bell" crotal from Hong Kong. Bronze, with Chinese characters on the side and attachment hole at the top. 3.125" long, 2.5" high. $75-100.

Luristan bell from ancient Persia. Bronze ovoid "birdcage" shape crotal with six unevenly sized openings and six holes in the base. The crotal pellet is approximately 3/4" in diameter and well rounded. 4" high. $250-300.

Old bronze Shinto temple bell from Japan. Crotal type bell, enclosed on bottom. 3.75" high. $115-135.

Oriental crotal bell, oblong in shape with four feet and handle. 7.25" high. $250-300.

The **gong**, another form of bell, has a long history as well. Gongs most likely originated in China, perhaps as early as the sixth century BC. They are usually flat and round, suspended from a framework which allows them to be struck with a hammer or mallet. A gong resonates most strongly at its center, in contrast to an open mouth bell whose rim produces the strongest vibrations. Gongs may also take the form of wide, cup or bowl shaped objects, often hung together in graduated sizes. Although they have an oriental heritage, gongs found their way east and were often used in the formal dining rooms of stately British homes.

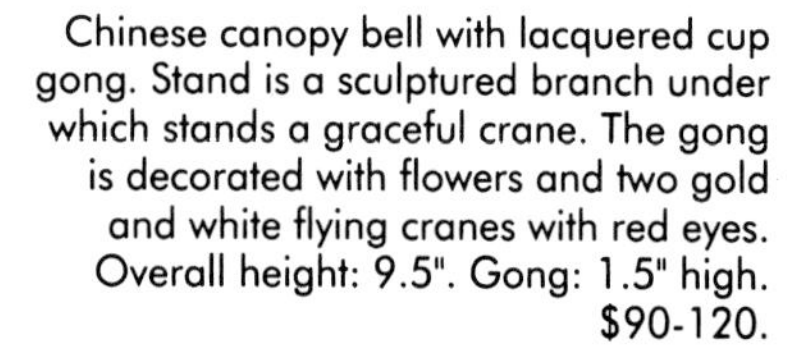

Chinese canopy bell with lacquered cup gong. Stand is a sculptured branch under which stands a graceful crane. The gong is decorated with flowers and two gold and white flying cranes with red eyes. Overall height: 9.5". Gong: 1.5" high. $90-120.

A second Chinese canopy bell, this one featuring three bowl gongs of graduated sizes suspended from a high framework with two monkeys on top. An acrobat lying on a platform balances the framework and gongs on the bottom of his feet. Overall height: 19.5". $275-350.

Oriental bowl gong with four gongs of graduated sizes attached to a green silk cord, each with a design of engraved dragons. From largest to smallest, the gongs measure: 3.25" high, 7" diameter; 2.5" high, 6" diameter; 2" high, 5" diameter; and 1.75" high, 4.25" diameter. $30-40.

Small gong mounted on a stand between two curved animal tusks, with holders in front for the wooden mallet. Most likely from Tibet. Stand: 11" long, 5" wide. Gong: 6" diameter. $100-150.

Brass gong with two tappers, one on each side. Ringing this English bell is accomplished by pushing down the button handles on either side; this action then moves the tappers against the gong. 5" high, 6" diameter. $100-125.

Ornate house gong on a walnut base. On the base are two prongs for holding a matching striker. This gong may have been from a home in St. Andrews, Scotland, where it matched the woodwork in the home; its tone is especially lovely. Overall height: 10". Gong: 5.75" diameter. $25-30.

The term **mechanical bell** is used to describe those bells in which a mechanical striking device of some kind makes the bell ring. Sound is produced by such actions as tapping or twisting a knob that sits on top, twirling an axle that holds the bell, snapping a protruding lever, or pushing on some part of the bell to activate a wound-up spring. The "ring" of a mechanical bell can vary from a pleasing chime-like tone to a somewhat strident buzz. Mechanical bells were often of a utilitarian nature, employed in hotels, restaurants, and shops to announce the arrival of a guest or customer. Others were strictly for amusement or decoration and occasionally made use of human or animal figures to operate the bell. Doorbells and bicycle bells are still other examples of this type of bell.

Double gong twirler bell with porcelain knob, dating from the late nineteenth century. 4" high. $45-60.

Mechanical type twirler bell of polished brass, American. The circular ball revolves on an axle which has knobs screwed onto each end. The bell is rung by twirling the knobs on the sides. Interior chime has two tones. 6.5" high. $40-50.

Brass snap bell. The striking mechanism is a horizontal arm on a spring with a hinged striker. This mechanical bell has a particularly fine ornate base with dark leaf fronds radiating out from the post to a burnished brass segmented circle. 5.25" high. $25-50.

Mechanical bell made in France, rung by twisting the handle. $50-75.

Spanish turtle bell made of damascene, shown from the side and top. Bell is wound up like a clock on the bottom; pushing the head or tail makes it ring. 6" long. $150-225.

Damascene snail bell, similar to the turtle bell but rings only by pushing the head (not the tail). 6.5" long. $250-400.

Unusual silver, chrome, and brass mechanical bell, purchased in London. The right side looks like a tree, its top comprising the bell part to strike. A large bird, perhaps a cockatoo, sits on top. The figure on the left may represent the man Friday from *Robinson Crusoe*. The switch on the front of the stand is not in working order, but is meant to control the mallet/striker held by the figure. 8" high. $200-250 if in mint condition.

Nickel-plated Bermuda carriage bell mounted on a small footstool. The bell is rung by stepping on the top of the stool. Plate attached to the stool reads: "Eagle Brand Best Carriage Bell." $75-100.

Finally, **chimes** may also be placed in the category of bells. These are sets of bells that are tuned musically, generally less than two octaves in range (tuned bells of twenty-three or more with higher ranges are known as carillons). Less standardized, but still considered bells, are the free-swinging chimes composed of objects that hang from a chain and knock against each other to produce a harmonious and musical sound.

Tubular chimes made by J. C. Deagan, Inc. of Chicago, Illinois, a company which also made plate chimes and cathedral chimes. This set has musical notes for military tunes, such as "Taps," "Reveille," "Church," and "Flag," embossed on the bars next to each chime. $100-150.

Armenian church chime with four graduated bronze chimes mounted beehive fashion on a polished wood base. There is a handle at the top, but no striker. 9.5" high, 8", 7", 6", 5" diameter chimes. $60-80.

Clay chimes in the shape of an Edwardian man. Stamped into the underside of the umbrella is the maker: "CLAY SOUP, Old Stage Road, Arrowsie, Me." 13" high. $15-25.

Chapter Two

Materials and Methods

One of the most interesting things about bells is the diversity of materials with which they can be fashioned. The material used certainly affects the overall style and appearance of the bell; conversely the impression or image desired by the bell's maker may dictate the type of material chosen. In some cases, the practicality of one material over another may override the more aesthetic aspects of a bell's creation. To illustrate the most typical kinds of materials used, this chapter provides a brief review of bells made from metal, glass, ceramic, enamel, and wood, including representative examples of each.

Metal

Bells can be constructed from many kinds of metal, including brass, bronze, copper, silver, and pewter. Some bells incorporate more than one metal or use a combination of metal and another substance. The surface of a bell constructed from metal can range from jagged and rough to satiny smooth. Some are polished to a lustrous shine, others engraved or inscribed with ornate decoration. On bells that incorporate human or animal figures, metal allows the artist or sculptor to carefully craft detailed textures, clothing, hair, and facial expressions, giving figures a sense of realism seldom duplicated with other materials.

Most metals used for bells are actually alloys, a mixture of more than one type of metal. Brass is generally an alloy of copper and zinc and is characterized by a yellow or gold hue, bronze is made of copper and tin and tends to be brownish in color, pewter is a combination of tin and lead (or other metals) and appears gray in color. As brass ages, it sometimes becomes darker and duller in color, making it difficult to accurately differentiate from bronze.

Metal bells are most typically produced through casting, a process in which the molten metal is poured into a mold and allowed to solidify. Sand and other materials are used to construct the mold. The Lost Wax method, an ancient form of casting also known as investment casting or *cire perdue*, is a painstaking, multi-step form of casting that produces particularly intricate and artistic pieces. In this method, first used more than five thousand years ago, a detailed wax model is first prepared, then covered with a ceramic shell. When the shell has set, it is placed in a kiln and fired, causing the wax to melt or vaporize out of the interior. After the shell is removed from the kiln, molten metal is immediately poured into the cavity left by the wax. Once the metal has cooled and hardened, the ceramic shell is removed, leaving a bell of exceptional quality and beauty.

Although we often think of silver as a "pure" metal, it too is alloyed (generally with copper) when used for the creation of artifacts and other decorative objects. The Brit-

Great Blue Heron bell. This heavy bronze bell depicting the head and neck of a great blue heron was cast using the Lost Wax process. Beak and eyes polished to a golden color, white wash added to the cheeks. Designed and made by John McCombie of Crooked Creek Gallery near Indiana, Pennsylvania. Marked "#29/75 JM96." 9" high. $200-250.

Heavy brass French ship's bell with ornate engraving, c. 1810. One side has the insignia of three bees (Napoleon's symbol), the other side has an engraving of the peace dove. Braided cord not original to bell. $75-100.

Detail of the three bees insignia.

ish reference to "sterling" signifies that the metal is 92.5 percent silver. Sterling silver can be used to coat, or plate, an object by employing one of two methods. Sheffield plate, used from about 1740-1860, involved heating and rolling sterling silver onto a base, usually of copper, which was then used to create the final object. With the electroplate method, patented about 1840 and still in use today, an already shaped metal object is coated with pure silver using a procedure known as electrolysis.

Three jester theme bells. Left: Brass bell with head of jester. 3.75" high. $35-40. Center: Bronze jester lies on his back on an ornate circular base with legs raised to hold up a twist bell painted green. The jester is a separate piece screwed onto the base. Under the base is inscribed "Ges: Gesch.," probably the name of the maker. 4.125" overall height. $130-160. Right: Heavy brass figurine of a jester or fool with a hobby horse and skirt which he is "riding." There is a partially obliterated scroll on the skirt and scrollwork around the hem. 4" high. $100-150.

Above and right: Florentine bell with underplate. Heavy cast brass or bronze bell and plate with elaborate relief ornamentation of scrolls, figures, faces, cherubs, and creatures. Many versions of this bell exist with different handles. Bell: 4.5" high. Plate: 5.75" diameter. $125-150.

Cast bronze deer's head with nose pointed up and the antlers curved back to one side. Modeled after an Iranian antelope-head rhyton (drinking cup). This bell is a copy of an original rhyton on display in the Louvre, Paris. 4.75" high. $50-70.

Crudely cast metal bell of unknown origin. Although the two heads at the base of the handle may represent jesters, this bell also bears a strong resemblance to one identified by Dorothy Anthony in *World of Bells No. 5* as a Nepalese export depicting a holy temple. 7.25" high. $30-50.

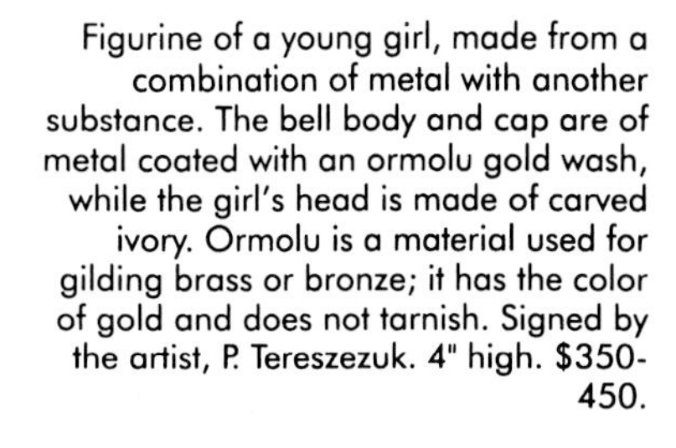

Figurine of a young girl, made from a combination of metal with another substance. The bell body and cap are of metal coated with an ormolu gold wash, while the girl's head is made of carved ivory. Ormolu is a material used for gilding brass or bronze; it has the color of gold and does not tarnish. Signed by the artist, P. Tereszezuk. 4" high. $350-450.

Milkmaid figurine bell. This figurine of a young girl with her left hand on her hip and her right hand carrying a pitcher is made of two different metals. The bottom appears to be of bronze, while above the waist the bell is made of cast pewter, washed slightly in a brassy color visible along the edges and depressions. 4.5" high. $50-60.

Detail of the bell's handle.

Metal lends itself well to various forms of ornamentation, two of which are illustrated here. Damascene, a technique originally used in Damascus, is produced by inlaying gold or silver wire on a more common metal such as iron, steel, or copper. This results in the kind of elaborate design seen on the back of the brass-bodied turtle bell. From Japan comes Chokin, an ancient art originally used to embellish the armor of twelfth century Samurai warriors, created by engraving a design on a copper base and then gilding it with precious gold and silver. Two examples of Chokin art, one with a fire-spitting dragon and one with a delicate bird and flower design, are shown below.

The shell top of this mechanical turtle bell from Spain is made of damascene, the body underneath of bright shiny brass. The fictional character of Don Quixote riding a horse, along with his assistant nearby crouching behind a donkey, are depicted on the shell. A key underneath the body winds up the ringing mechanism; either the turtle's head or tail can be pressed to make it ring. 5.5" long, 3" across. $150-225.

Chokin decorated bell with hummingbird design. 5.5" high. $4-6.

Chokin dragon bell, with 18 kt gold electroplate finish on a plain gold base and a flat, scroll-edged handle. Set into the handle is an engraving of a dragon facing a flaming pearl ball. Incised on back: "ORIGINAL CHOKIN COLLECTION, DYNASTY, JAPAN." 4.25" high. $15-20.

Silver bell done in repoussé style, in which the scenes of Holland on the base are impressed into the metal from the back. The arms of the windmill handle turn. 4" high. $60-75.

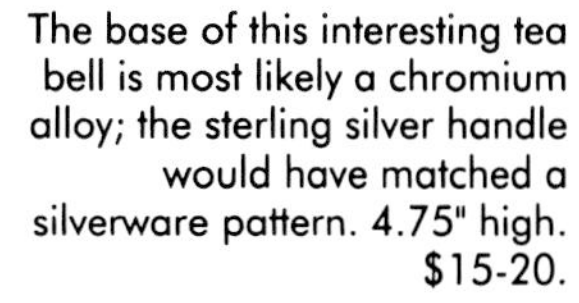

The base of this interesting tea bell is most likely a chromium alloy; the sterling silver handle would have matched a silverware pattern. 4.75" high. $15-20.

Combination glass and silver bell. The top of the bell unscrews and the silver filigree overlay can be removed for cleaning. 4.75" high. $100-125.

Gorham silver-plated tea bell with narrow stem handle ending in a teardrop shaped top which looks as though it is "wrapped" in silver coils. 3.75" high. $35-55.

Heavy metal bell of silver appearance with exceptionally fine tone, made by the Wm. Chase Company. The Art Deco handle has a plain stem surmounted by a black ball of ebony appearance. Inside the bell is a hallmark of a rearing centaur with bow and arrow, and to one side "CHASE, USA." 3" high. $20-30.

Silver-plated bell made by the Danbury Mint, part of a series of songbird bells. This one features an Eastern bluebird, which appears to be made of pewter. 3.5" high. $20-25.

Detail of St. James figure, showing the hat and cape.

Plain, silver-plated bell base, surmounted by a cast pewter figure of St. James clad in traditional pilgrim's garb with a book and a staff. His hat and cape are marked with the "cockle" shell symbol of the apostle, and he is standing in his bare feet on a shell. 4.5" high. $35-50.

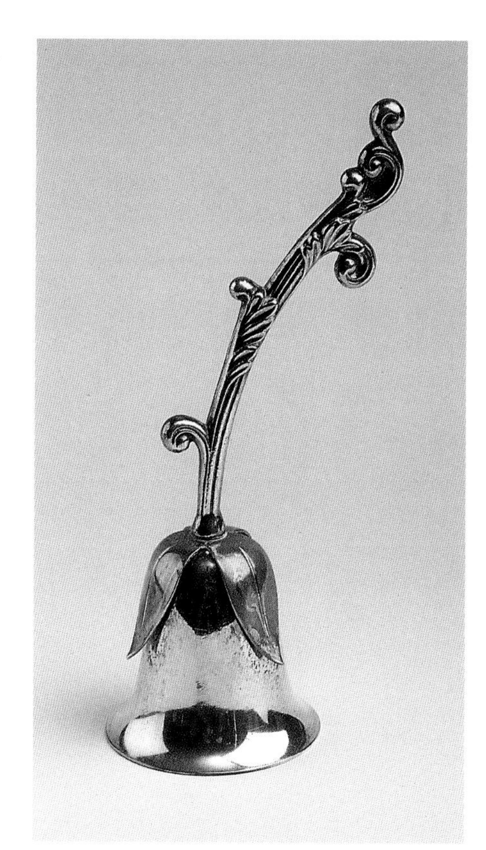

Mixed metal "flower" bell with curved stem. The chrome or silver-plated bell base has three brass "petals" folded over the top and shoulder. The handle is a graceful, stylized "stem" made of silver with curlicues suggesting leaves. Purchased in Brazil, c. 1920s. 4.625" high. $15-25.

The Bovano Company of Cheshire, Connecticut, manufacturer of these delicate copper and enamel bells, advertises that the flowers will not fade, even in direct sunlight! Flowers include lily of the valley, daisy, violet, aster. Note that the daisy bell, while not specifically identified as a Bovano bell, is believed to be of their manufacture due to its similarity to the others. $10-20 each.

Glass

Clear or colored, glass bells have few rivals when it comes to elegance and grace. Even a modestly priced, mass produced glass bell takes on an aura of enchantment when placed near a window or mirror, light gently dancing off its smooth contours.

Despite variations in size, hue, fragility, and decoration, glass bells generally start with the same basic elements. Essentially, glass is created by the high temperature fusion of silica (sand, quartz, or flint) with flux (an alkali material made of soda, potash, or both). Other ingredients, such as lime, alumina, or lead oxide, may be added to produce special effects. Crystal, for example, must contain at least 24 percent lead oxide; this formula gives the crystal its characteristic clarity and brilliance. Another name for lead glass is flint glass.

To create glass objects, hot, liquid glass is given form in a number of ways, including hand blowing, mold blowing, pressed glass, and machine-made techniques. The glass is then slowly cooled to protect its integrity, after which patterns can be added through such processes as cutting, engraving, or etching.

While it is beyond the scope or intent of this book to provide a history of glassmaking or to review the many noted companies which produced fine quality glasswares, the array of glass bells presented here illustrates a variety of popular styles and techniques. Bells from some of the leading companies are included, such as Fenton, Pairpoint, Imperial, and Steuben, yet this is certainly but a small sampling of the many collectible glass bells available.

Early European glass bells were primarily colorless; the advent of beautifully colored glass bells began with the Bohemian invention of overlay glass in the early nineteenth century. This process involved multiple layers of glass; the outer layer is usually milk white and is cut in decorative patterns to show the colored layers underneath. (Springer 1972, 88)

Italian glassmakers from the island of Murano are renowned for their skill in creating several types of elaborate glasswares. Bells made of Millefiore ("thousand flowers") glass are covered with multi-hued stylized flowers, aptly living up to their equally colorful name. To obtain this effect, tiny pieces of colored glass are applied to the basic glass shape while it is still hot and workable. Continued heating and working results in the original glass and the tiny pieces merging into one layer. (Kleven 1989, K) Ribbon glass bells, also from Murano, feature strips of colored glass applied to the basic glass layer, some with a twisted appearance for added interest.

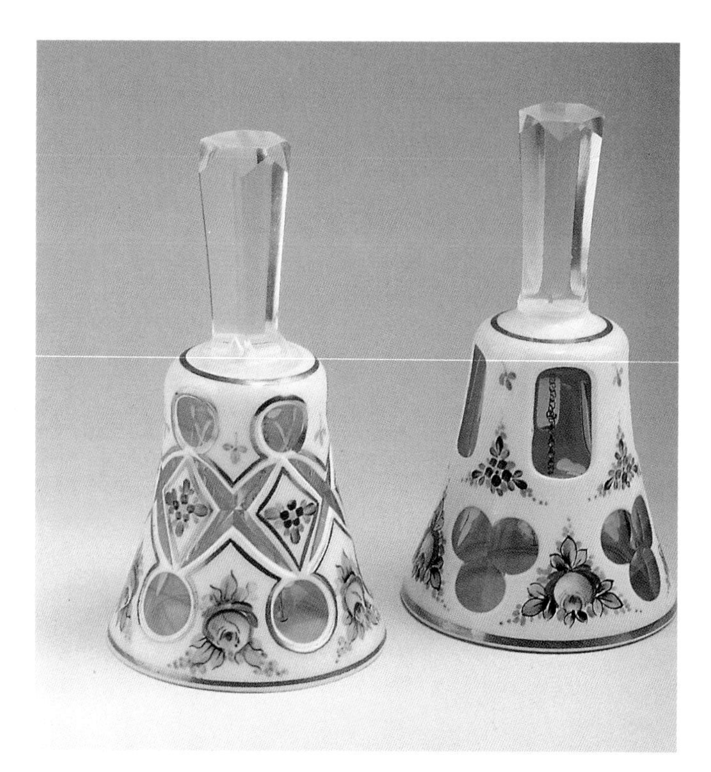

Two Bohemian overlay glass bells from Czechoslovakia. The base of the bell on the left is thick, white opaline glass into which cuts have been made to expose the transparent green layer underneath. The bell on the right is made of pale blue glass with a white overlay and four rectangular "windows." The outer white layer of both bells is painted with pink roses and green leaves. Both 5.25" high. $40-60 each.

Bohemian glass bell of hand cut lead crystal. 7" high. $20-25.

Blue overlay glass bell with clear crystal handle. 5" high. $35-50.

Two Bohemian glass bells with engraved designs and clear glass handles. Left: Deep ruby, 5.75" high. Right: Deep amber, 4.5" high. $20-25 each.

Two bells made of Italian Millefiori glass, one a cowbell shape and the other an angel with clear glass head and wings. Cowbell shape: 6.25" high. $20-30. Angel: 5.25" high. $30-40.

Millefiori glass bell with delicate red bird finial. 3.25" high. $30-40.

These two contemporary ribbon glass bells are from Murano, Italy. The bell on the left has multi-color "ribbons," a scalloped lip edge, and a clear loop glass handle. Approximately 6" high. On the right is a glass bell with alternating vertical panels of white lattice pattern stripes and twisted pink "ribbons" set into clear glass. 4.5" high. $20-25 each.

France is the origin of a rare and highly sought series of delicate flint glass bells, each featuring enameled bronze animals as handles and coordinating clappers that relate cleverly to the animal on top. Other than their French ancestry, however, little is known about these small, intriguing bells. They are believed to have been manufactured in the late nineteenth century in an unknown factory, possibly the Baccarat factory in Lorraine. Standing approximately 4" high, the flint glass bells are primarily found in green or amethyst and some have the word "France" written in black on a lower edge of the bell.

Flint glass bell, cat handle with mouse clapper. $500-900.

Flint glass bell, rooster handle with egg and claw clapper. $500-900.

Flint glass bell, bear handle with pine cone clapper. $500-900.

Flint glass bell, pig handle with acorn clapper. $500-900.

Flint glass bell, poodle handle with sugar cube clapper. $500-900.

Flint glass bell, swan handle with metal ball clapper. $500-900.

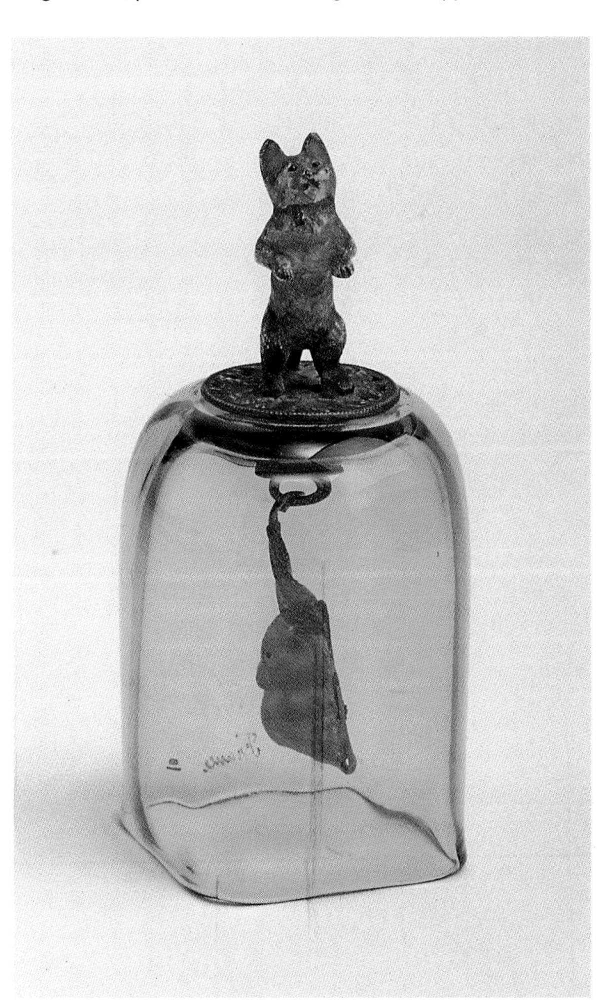

Flint glass bell, cat handle with mouse clapper. $500-900.

Flint glass bell, rabbit handle with carrot clapper. 4" high. $500-900.

Three of the enigmatic flint glass bells shown as a group.

Not so much a mystery as a disagreement surrounds the large glass bells, mostly made in England, which have become known as wedding bells. Some feel these bells were made as commercial products by glass manufacturers in the towns of Bristol and Nailsea, others believe the bells were actually "whimseys," after-hours experiments created to suit the fancy and show off the skill of individual artisans at the glass factories. Regardless, most bell fanciers do agree that these creations are among the grandest and most spectacular of all glass bells. Their handles, generally tall steeples ending with one or more gradually tapering knobs, required exceptional skill to master. These bells (shown on pages 31-33) are found in shades of red, blue, green, and other colors, some with swirls or stripes embellishing the base. Their name comes from the belief that they were given in celebration of marriage, but this is now felt to be of dubious truth. In *World of Bells No. 5,* Dorothy Anthony notes that "although some were given as wedding gifts during Queen Victoria's reign, 'wedding bell' is a misnomer and we should designate them English glass bells." (Anthony N.p., n.d., Plate 5) To distinguish these bells here, we will refer to them as "wedding-type" bells.

Reverse painted bell. This simple glass bell has a detailed winter scene painted on the *inside*. The artist used delicate brush strokes to paint the inside of the bell through a small opening. The process involves painting in reverse, that is, the details (such as the horse's harness) are painted first, then the background. The bell has no clapper, as this would disturb the paint on the inside. Manufactured by Chase International. 3" high. $15-25.

Three glass bells made by the Fenton Art Glass Company. From left: Ruffled edge bell with clear faceted bead clapper hung on a silver chain from the traditional Fenton twin fingers, 6.5" high; milk glass bell with ruffled daisy and button pattern, purchased as a second at the Fenton Company, 6" high; hobnail milk glass bell, 5.75" high. $15-20 each.

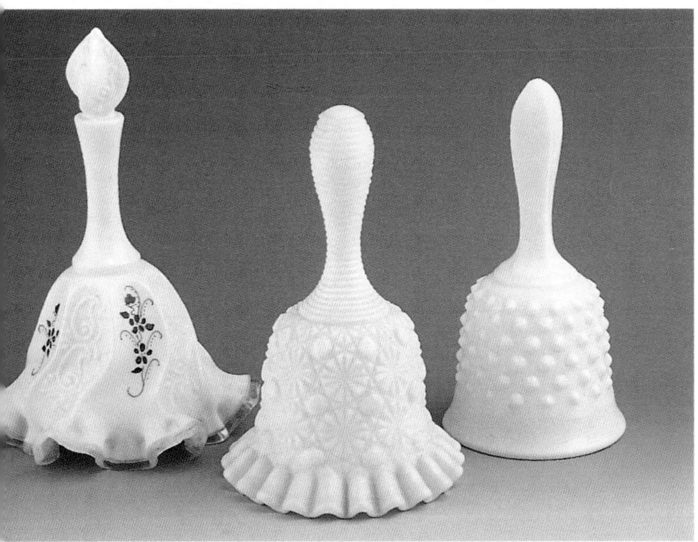

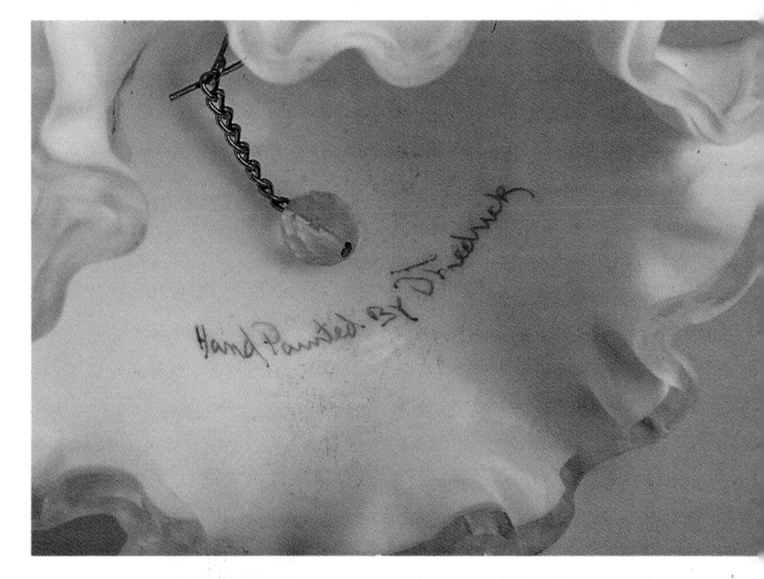

Close-up of the clapper and artist's signature inside the Fenton ruffled edge bell.

Fenton hobnail bells in blue and amberina. Both 5.75" high. $15-20 each.

Crystal bell from the Steuben Glass Company in Corning, New York. To make the highly desirable air twist handle, air bubbles are captured in the unshaped molten crystal, then pulled lengthwise and twisted in combination with the glass. Clapper inscribed with the Steuben signature. 6" high. $150-175.

Lead crystal bell, hand cut, in cobalt blue. The handle is a slender six-sided pillar, gradually diminishing in circumference from bottom to top and notched along each edge. 6.625" high. $25-35.

White wedding-type bell. The opaque white base has a clear glass handle that is cemented to the base with white plaster in which two small rings are embedded. The rings hold a metal link, which in turn holds a silver chain with a clear glass bead. This clapper looks newer than the bell itself and may have been a replacement for the original. 11.5" high. $250-300.

Pink wedding-type bell. $125-150.

Wedding-type bell of deep cobalt blue with swirl pattern on the glass. Handle of clear glass with typical steeple finial. 10.25" high. $150-175.

The three above bells displayed together.

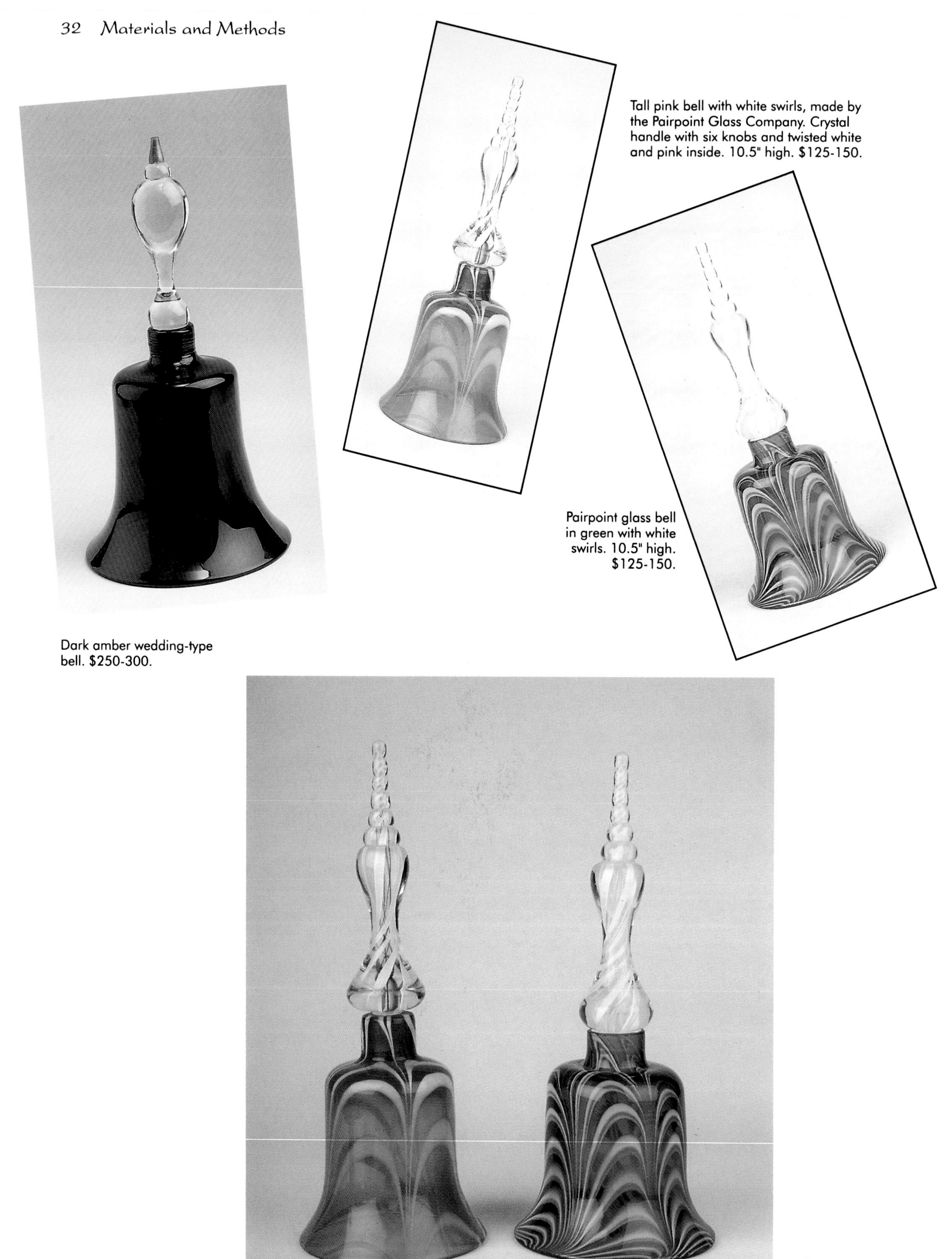

Tall pink bell with white swirls, made by the Pairpoint Glass Company. Crystal handle with six knobs and twisted white and pink inside. 10.5" high. $125-150.

Pairpoint glass bell in green with white swirls. 10.5" high. $125-150.

Dark amber wedding-type bell. $250-300.

The two Pairpoint bells together.

Wedding-type bell with five knobs on handle. 9.25" high. $250-300.

Wedding-type glass bell by unknown manufacturer, probably English. Large pink bowl with spiral white striping and blue bulbous handle with three knobs at the very top. 13.25" high. $250-300.

The tall steeple handles of these glass bells clearly illustrate the skill and artistry required to create them.

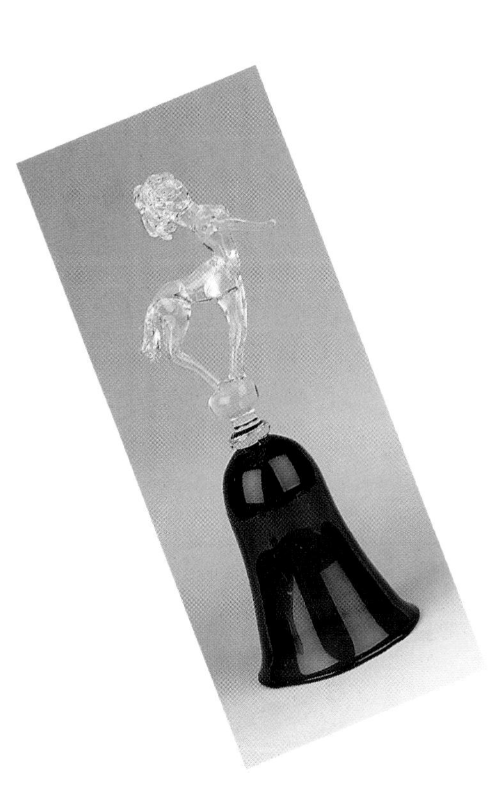

Two cobalt blue glass bells with crystal figural handles. Hand blown glass, crystal clappers. Approximately 6.5" high. $10-15 each.

The contrast of the applied flower decoration in gold against the blue glass bell makes this an attractive accent for the home. 6.25" high. $10-15.

Three examples of marble glass bells from the Imperial Glass Company. Marble glass (sometimes referred to incorrectly as slag glass) used to be called "end-of-day" glass and was thought to be made from glass "left over" at the close of work. This material is now felt to be more accurately described as marble glass, the blending of which is a more difficult process. From left: Purple marble glass, 5.75" high; ruby marble glass, 5.625" high; jade marble glass, 5.5" high. $30-40 each.

Ceramic

A ceramic object is one that has been made from clay fired in a kiln. The two basic types of ceramic are earthenware, also called pottery, and porcelain, also called china. They differ in several important ways.

Earthenware, the older of the two types with a history dating back to early Egyptian times, is fired at a relatively low temperature, is opaque, and is porous. It must be glazed in order to be waterproof. Porcelain, originating in China around the ninth century AD, is fired at a very high temperature, is translucent, and is non-porous. It is waterproof even when unglazed.

Glazing is a form of decoration that gives ceramics its well-known smooth and shiny finish. Earthenware or porcelain objects that are fired but not glazed are referred to as biscuit, or bisque. When glazing *is* desired, it can be done either before or after the piece is fired. If applied after the firing, a second firing is required in order to fix the glaze.

Although well developed in China for many years, porcelain was not manufactured in Europe until the early eighteenth century. Italy and France did produce a very small quantity of what is called artificial or soft-paste porcelain several hundred years earlier, however the first true or hard-paste European porcelain manufacture is generally credited to the Meissen Company of Germany.

Once the manufacture of European porcelain was perfected, porcelain articles of great diversity reached unprecedented popularity, a popularity which continues today. Bells made from porcelain or earthenware, like other ceramic objects, vary considerably in terms of their shape, finish, and decoration. Many come from such distinguished porcelain companies as Wedgwood, Limoges, Hummel, Royal Bayreuth, Royal Tara, Herend, Belleek, Coalport, Lenox, and the aforementioned Meissen; the experience and history of these companies is reflected in the superior quality of their bells. Other ceramic bells may have a less prominent origin, yet can still be appreciated for their design, texture, and style, and may be more practical for beginning collectors to seek and acquire. Pottery, too, is a favorite medium of many talented and creative artists who delight in producing bells that are unique and exceptionally pleasing.

Many books and materials are available for those interested in delving further into the manufacture and types of ceramics, a subject which easily fills many pages. In addition to the ceramic bells shown here, the reader will find both earthenware and porcelain bells throughout other sections of this book.

Glazed porcelain bell with applied flowers, by Meissen. Gold bands above lip, around base of handle, around widest part of handle, above and below pink band on handle, and on top of handle. These features, along with the Meissen blue crossed swords trademark and wooden ball clapper on rod are common to most Meissen bells. 4.5" high. $175-200.

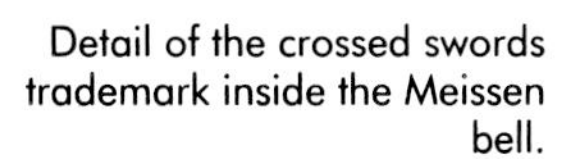

Detail of the crossed swords trademark inside the Meissen bell.

Three Irish bells of Belleek porcelain decorated with holly, lily of the valley, and shamrocks. Average height: 5". $50-75 each.

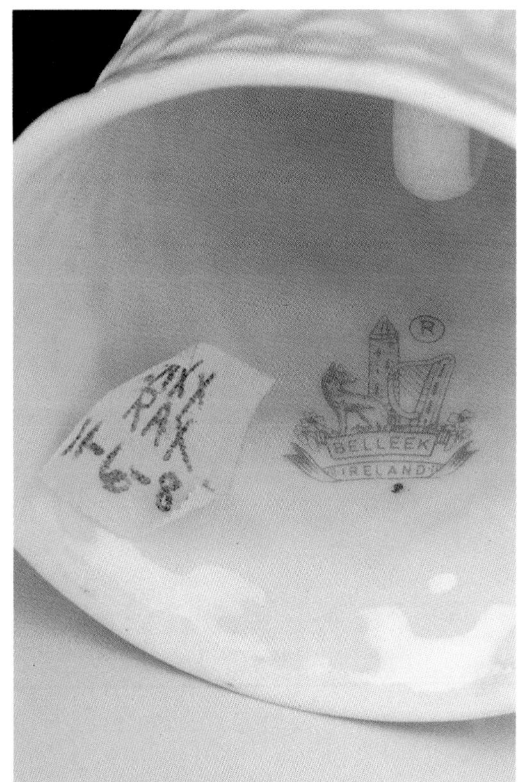

Belleek manufacturer's mark found on the holly bell, printed in gold.

Irish Belleek bell with harp shaped handle, decorated with shamrocks. 4.375" high. $50-75.

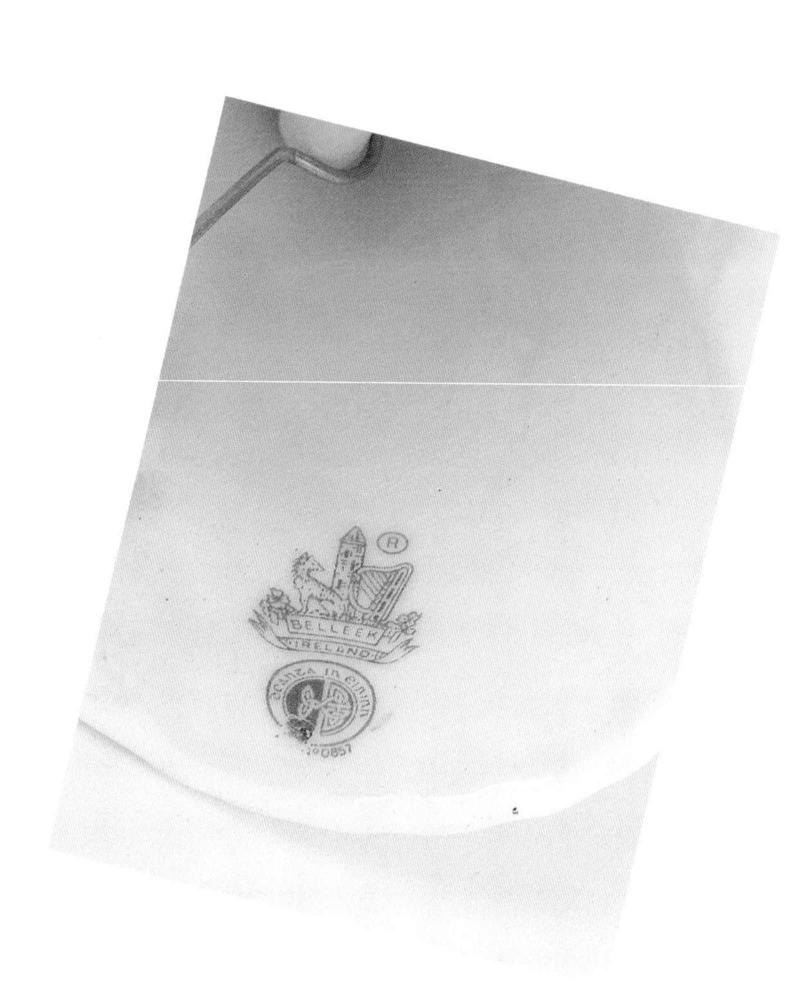

Belleek manufacturer's mark found on the harp bell.

Wedgwood bell of blue and white jasper with three groupings of penguins, all different, standing on ice floes. Incised on the white rim around the base is: "New Year 1979." 3" high. $25-30.

Wedgwood Four Seasons bells in light blue and pink jasper with designs in white relief illustrating the four seasons. Clappers are suspended from a jasper ring upon which is impressed "WEDGWOOD ENGLAND." Both 4.25" high. $60-70 each.

Bagpipe bell of Quimper pottery. Glazed pottery in "bagpipe" shape with characteristic Quimper painting of a French peasant on one side and flowers on the other. 3.5" high. $40-50.

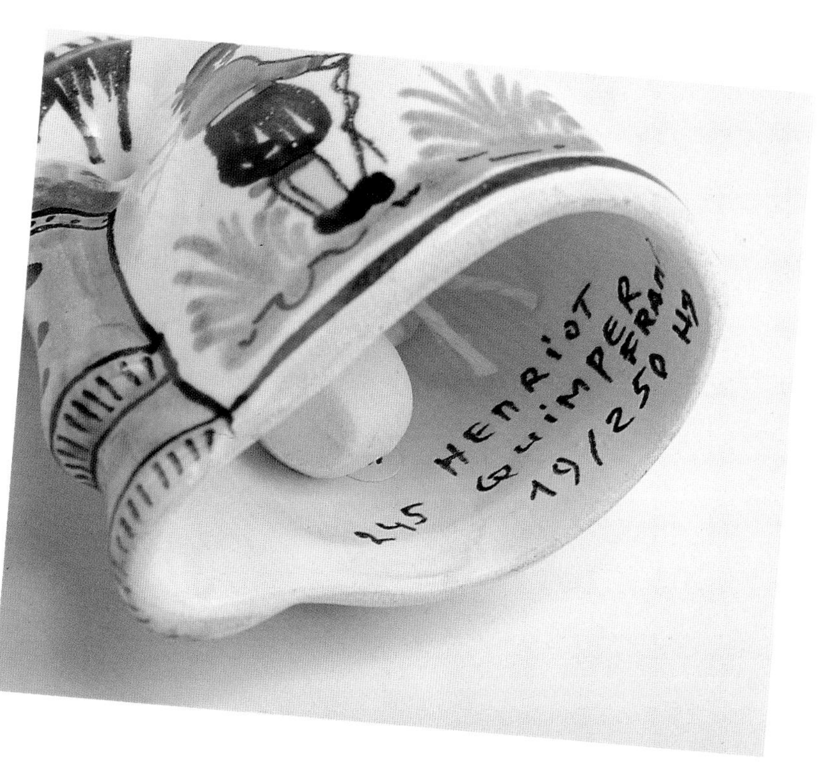

Detail of the mark inside the Quimper bell.

Coalport white bone china bell, made in England, with green leaf and flower decoration around the matching green handle. The Coalport Company is particularly known for its skill in producing flowers on china. 3.5" high. $20-25.

White ceramic bell with heart shaped handle and delicate rose design on front. 4.5" high. $10-15.

Cowbell shaped ceramic bell with gold handle, made in Germany. 4.25" high. $40-50.

Dainty Limoges bell with wooden handle. 4" high. $5-10.

Nippon china bell with hand painted roses and gold handle. 4.25" high. $10-15.

Tapestry design ceramic bell, probably made by the Royal Bayreuth Company of Germany in the late nineteenth century. The unusual finish on such tapestry bells was obtained by covering the bell with a coarse cloth prior to the firing process. The heat of the firing burned the cloth away, resulting in the tapestry-like effect. This bell has a wooden clapper, also unusual (though in the tradition of Meissen bells) because it is a different material than the bell itself. 3.25" high. $225-250.

The exterior texture of the bell resembles needlepoint tapestry.

Three bells of spaghetti ware, a type of ceramic that comes from various countries. The uppermost bell is from Italy and is 3.5" high; the smaller pink bell on the bottom right is from Japan and is 3" high. The smallest bell, origin unknown, is only 2.125" high. $15-20 each.

White spaghetti ware bell with applied colored flowers. 2.5" high. $50-65.

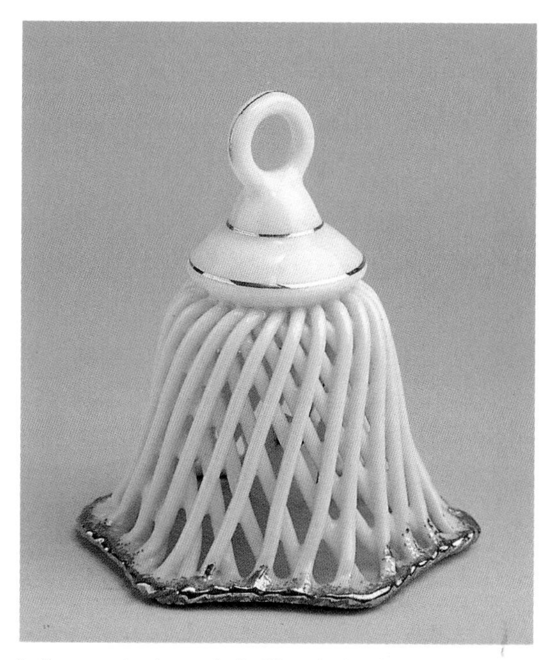

Italian spaghetti ware bell. Glazed white bell with gold on upper side of the braided hexagonal lip and gold lines around the shoulder and handle. The rounded strands are indicative of a hand made bell, while those with flat strands are most often machine made. 2.75" high. $25-30.

Spaghetti ware bell with large, three-dimensional applied flowers. 4" high. $5-10.

The English cottage bell on the left, with its thatched roof, leaded windows, and pink hollyhocks, is reminiscent of the Shakespearean countryside. Hand painted ceramic from Staffordshire, England. 4.75" high. The bell on the right is fashioned like a brick house, with the handle serving as the "chimney." Glazed ceramic, no clapper. 4.5" high. $25-40 each.

Glazed white china bell with transfer pattern Australian platypus in natural colors. This bell was issued by *Australian Geographic* and is so marked. 4.25" high. $25-30.

Left: Norman Rockwell's "Missing the Dance" is the design on the front of this glazed ceramic bell. 6.25" high. $5-10.

A Currier and Ives winter scene decorates this blue and white china bell with a scalloped rim. 5.5" high. $2-3.

Enamel

The process of enameling involves heating a paste made of powdered glass and water to a high temperature and then fusing it onto a metal surface such as copper, bronze, silver, or gold. Several of the techniques used to separate and define the enamels are illustrated on the bells shown here. Additional enameled bells can be seen in Chapters Three and Four.

Although the word cloisonné is French, it is the Chinese who are world renowned for this type of enameling. To make a cloisonné bell or other object, thin metal wires are attached to the metal base. The compartments formed by these wires are known as cloisons (French for "cell"); these are filled in with the enamel paste in colors appropriate for the design and then fired at a high temperature. The resulting decoration remains permanently bonded to the base. Oriental cloisonné used copper for both the base and the wires and the piece was often fired multiple times to raise the height of the enameling higher than that of the surrounding cloisons. (Hammond 1976, 2)

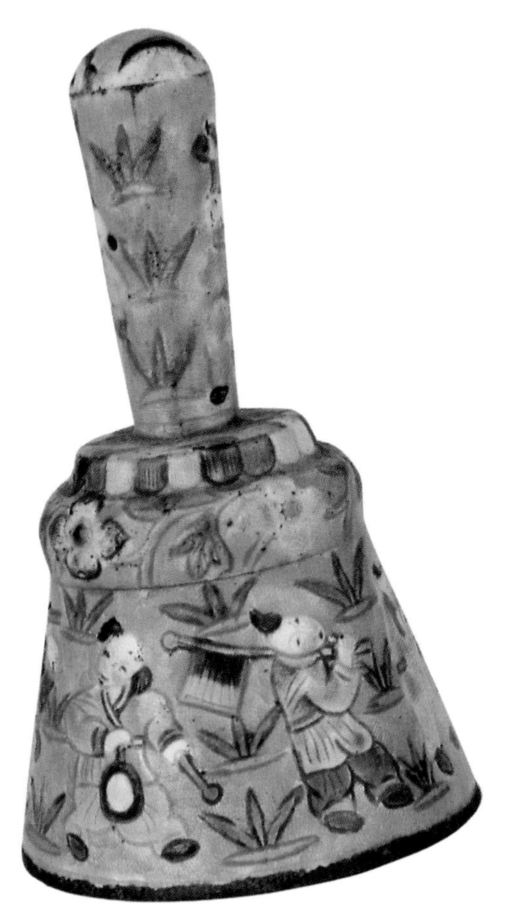

Chinese enameled bell decorated with the Seven Gods of Happiness. 4.75" high. $35-40.

Champlevé is a technique similar to cloisonné, the difference being that the compartments to receive the enameling are excavated out of the metal itself, rather than being built up upon it. The excavation is accomplished via chiseling, routing, carving, engraving, grinding, or acid etching the metal; in some cases wires are attached to the cut out areas creating a combination cloisonné and champlevé effect. (Hammond 1976, 2)

Repoussé enameled bell on a copper-plated, silver base. The bare chested man with upraised arms on the handle depicts Hotei, the Chinese god of good fortune. 5" high. $80-100.

In repoussé enameling, the underside of the object being decorated is used to prepare it for the enamel paste. The design or decoration is embossed into the metal from underneath and the pushed-up areas created by the embossing function the same way as cloisons. Placed into the depressions between the pushed-up areas, the enamel follows the contours of the depressions and the finished piece requires very little grinding or polishing as a result. (Hammond 1976, 3)

Plique-á-jour, the final type of enameling shown here, has been used in France and Italy but was most successfully employed in Russia during the late nineteenth century. Translating to "brilliant light of day," plique-á-jour enameling lacks a metal backing and the pieces tend to be extremely fragile as a result. The overall effect is reminiscent of stained glass windows. (Hammond 1976, 5)

Repoussé enameled bell from China, very lightweight. Center band depicts the Eight Trigrams of Fu Hsi, metaphysical symbols found in the classic *I-Ching*, or "Book of Divination." The trigrams represent the eight possible permutations created by three lines placed one beneath the other. Note eyes peering out from the bottom band of this bell. 4.625" high. $60-75.

Champlevé enameled bell from France, heavy brass with multi-color pattern. 3.5" high. $250-350.

Three bells illustrating a variety of enameling techniques. Left: White cloisonné bell with gold handle simulating bamboo and an applied three dimensional flower on the center of the crown. 3.125" high. Center: Plique-á-jour enameled bell using a combination of transparent and translucent enamel. 2" high. Right: Gold based cloisonné bell. The polished cloisonné surface stands out in relief against the rougher gold bell surface. 3" high. $25-30 each.

Wood and Related Materials

Perhaps less common but no less attractive, wooden bells can be painted in bright colors or left in their natural state, sanded and buffed to a soft sheen. Any number of different woods can be used, as documented by the assortment shown here.

Related materials used for bells include papier mâché, a material indirectly made from wood, and bamboo, the product of a tree-like tropical plant.

Wooden bells made by Carl Hebel of Henderson, Minnesota, with heights ranging from 2.5" to 5". From left: Spruce, maple (smallest), myrtlewood (tallest), basswood, cottonwood. $15-30 each.

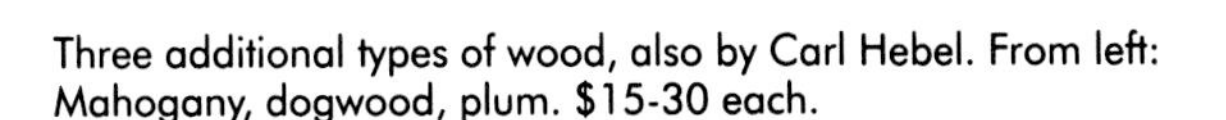

Three additional types of wood, also by Carl Hebel. From left: Mahogany, dogwood, plum. $15-30 each.

Two wooden bells, both made by artists Cal and June Brockman of Shawnee, Wisconsin. Left: Turned wooden bell painted light tan with a Rosmaling flower painting on the bell base, initialed JGB. 6" high. Right: Lathe turned, aromatic red cedar bell with smooth finish. Inside left unpolished to preserve the aroma. 2.75" high. $8-20 each.

These painted wooden bells come from the island of Java. A large orange carrot with green leaves at the bottom serves as the clapper for the rabbit's head, while a long clapper hanging below the whiskered nose simulates a tongue on the cat's head. The clapper on the red fish bell consists of two wooden rods suspended from a string fastened through holes on the side. Rabbit: 4.5" high; Cat: 3" high; Fish: 5" high. $10-15 each.

Small bamboo bell in shades of green, black and gold with metal interior. Purchased in Guilin, China. 2.75" high. $6-8.

Assortment of papier mâché bells, all richly decorated in bright colors and intricate designs. From left: Slender bell purchased in Victoria, Canada, 3.5" high; black lacquered bell with gold swirl background, purchased in China, 3.5" high; ivory bell with red and yellow flowers, made in Kashmir, India, 2.5" high; red and green hand painted bell also made in Kashmir, India, 3.25" high. $6-8 each.

Chapter Three

Globetrotting—Bells From Around the World

"Because bells speak the fundamental language of music, they speak all languages," writes Jane Yolen in her book *Ring Out!* And it is true—no cultural or geographical boundaries limit the world of bells; they are found and enjoyed worldwide.

This chapter presents a sampling of bells spanning seven continents—Europe, Asia, Africa, Australia, North and South America, and Antarctica—illustrating both the diversity and the concomitant universality of bells around the world. The sampling is by no means all-inclusive, and bells representing different countries will also be found in subsequent chapters. In Chapter Four, for example, additional Asian bells are featured in the section on religion and a host of international bells are pictured in the section on animal bells.

Additional faïence figurines in native costumes from the provinces of Aurillac and Alsace. 8" high. $50-75 each.

Glazed faïence pottery bells from France depicting costumes of the French provinces. Left: French peasant woman. Printed inside is: "MADE IN FRANCE 1970." 8.25" high. Right: Peasant woman from the Alsace province. Her headdress is a unique feature of this region. 6.5" high. $50-90 each.

Close-up of mark inside the Limoges bell.

French bell made of Limoges china. Green handle and base, painted roses on body. Marked inside: "ELITE Limoges, France." 5" high. $40-50.

Left: Green Wedgwood bell from England depicting white cherubs in various agricultural pursuits. 4.5" high. $60-70.

Right: Unglazed bisque figurine bell from Ireland. 4.5" high. $3-6.

Bone china bells made by the Royal Tara Company of Ireland. Left: Cowbell shaped bell featuring Celtic design of the letter "G" from the *Book of Kells*, an eighth century illuminated manuscript. Right: Royal Tara children's bell. Both 5" high. $20-40 each.

Close-up of the trademark and clapper from the Royal Tara children's bell.

Hand painted ceramic bell from Holland with blue jay handle shows a Dutch scene on one side, floral decoration on the other. 9" high. $15-20.

Three bells from Spain by Lladro. The 1987 and 1988 annual Christmas bells are shown on either side of a graceful figurine depicting a woman holding an umbrella. Christmas bells: 3" high. $25-30 each. Figurine: 5" high. $125-150.

Several of the bells from our "mini-tour" merit special attention. While most of the European bells can be clearly ascribed to a specific country, others are of a less definitive origin. One of probable German derivation is "cowbell" shaped, with its front and back surfaces richly decorated in bas relief. Circular bust portraits surrounded by scrolls and leaves appear on each side. Various opinions have been put forth about who these portraits represent but general agreement holds that the male figure is a Teutonic knight on a German bell.

Heavy bronze bell with circular portraits on either side. Of indefinite origin, most likely Teutonic. 3" high. $60-80.

Czechoslovakian overlay glass bell with painted colonial dancers. Crystal handle and gold trim. 6" high. $25-35.

Porcelain bell from Budapest, Hungary, made by Herend. The shape of this dainty bell is typical for this company, although they do make other shapes as well. 2.75" high. $60-75.

The mark and clapper inside the Herend bell.

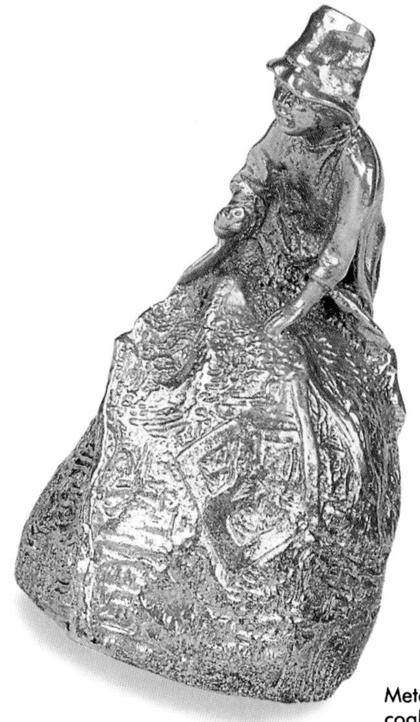

Metal bell with Russian boy on a coal pile, a reproduction of an earlier bell. Marked inside as made in Russia. 4" high. $50-75.

Earth-toned ceramic bell from the small town of Derutta, Italy, not far from Rome. 3" high. $3-5.

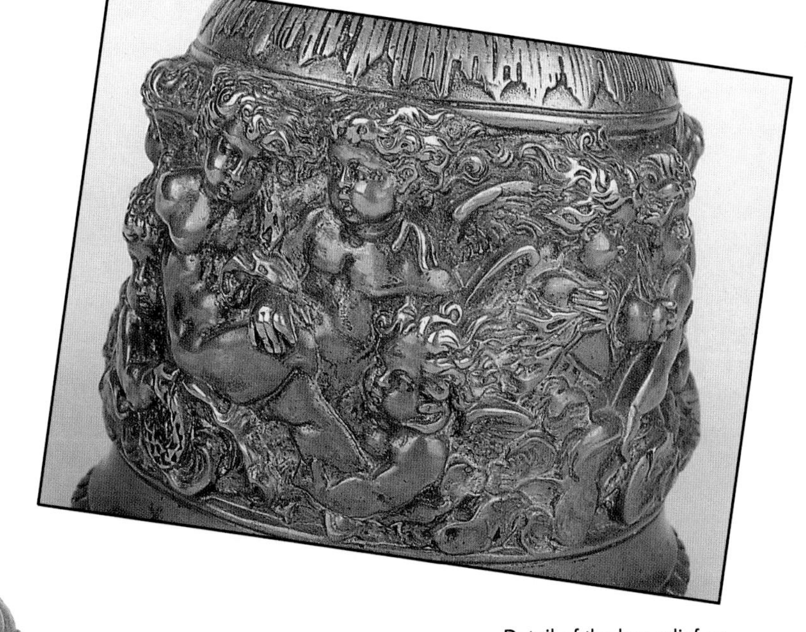

Detail of the bas relief on the Renaissance bell.

Large, heavy Italian bell with bas relief of cherubs around the base, most likely dating from the Renaissance period. 8.5" high. $150-200.

Capo di Monte bell from Italy, of glazed ceramic. These popular bells were named for the Italian city where they were made. The bell's main feature is a pair of angelic cherubs both front and back in a 3" diameter frieze. Marked inside in gold lettering: "1535/366 Italy." 4.25" high. $15-20.

The Capo di Monte bell interior, showing the mark and gold lettering.

Even greater mystery surrounds a group of metal bells commonly referred to as Hemony bells. Found in various shapes and sizes, some with unusual and unidentified figures, the common feature of these bells is an inscription around the bottom that reads: *F. HEMONY ME FECIT ANNO 1569*, Latin for "F. Hemony cast me in the year 1569." While the Hemony brothers, Franz and Pieter, were well-known early bell casters from the Netherlands, the date from the inscription does not make sense given that they lived and worked in the seventeenth century, not the sixteenth. As Lois Springer notes, the possibility that the Hemonys simply inscribed the wrong date has generally been discarded; being a "notedly conscientious workman, ...[Franz] would never have permitted this to escape his attention." (Springer 1972, 94) No answer to this enigma has yet emerged, although many believe that the bells were actually cast by later bellmakers seeking to exploit the Hemony brothers' name who were careless enough to use the incorrect date on their reproductions. Variations found in the inscription and dates are equally perplexing, although they may lend credence to the idea that the bells were actually cast by a series of bellmakers. One variation bears the inscription *JACOVES SERKE HEFT MY GHEGOTEN Aº 1370* ("Jacob Serke has cast me in the year 1370"); bells with this inscription are generally referred to as Serke bells.

Brass Hemony bell, very heavy. A Russian bear on the handle holds a shield with what appears to be the Russian emblem of arm-and-hammer. The scene around the base of the bell depicts jousting knights on horseback and a man leading a hound. The date on this bell's inscription is 1565. 8" high. $250-300.

Right: This bronze "passport" bell from Japan was made by Mr. Kashwiagi, a member of the American Bell Association's Japanese chapter. Such bells were originally used in Japan around the seventh century as an ancient type of identification system for official messengers traveling throughout the country. 4" high. $10-25.

Another Hemony bell, this one with figural handle of an unidentified haggard old person. $80-100.

Set of two brass "Goddess of Happiness" bells from Japan. Symbols on the back translate to "luck" and "happiness." 2.5" high. $5-8 each.

Two bronze bells from Japan. Left: Dohtaku bell, with panel of six different pictures front and back. An original version of this bell, c. 100-300 AD, is housed in a museum in Kobe, Japan. Dohtaku bells have a characteristic oval shape with fins on either side and are still sometimes found in ancient Japanese burial mounds. Their exact significance is not known. Right: bronze replica of large Japanese bell. Green patina with Japanese lettering on the side. $25-40 each.

Left: Japanese lady bell with a particularly melodious sound, made of paper glued to a cylindrical shape. A label with Japanese letters is on the side. $5-8.

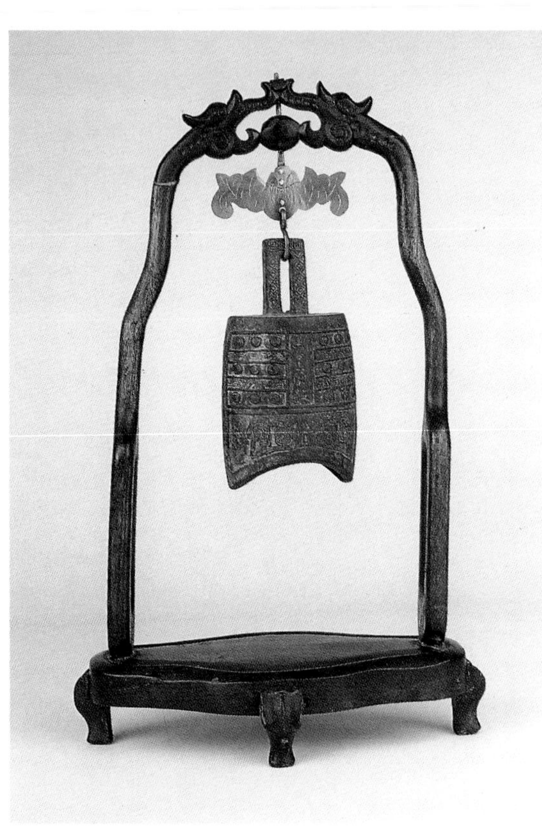

Right: Chinese temple gong on wooden stand. Overall height: 15.5". Gong: 4" high. $20-25.

These small keg-shaped bells are known as "The Lucky Hammers." The printing on the front of the box stands for "eternal happy life." 1.25" high. $5-8 each.

Here are "The Lucky Fish," a pair of Japanese fish crotals similar to ones that can be seen on temples in Osaka. 3" high. $5-8 each.

Reproduction Ming Dynasty bell with matching openwork stand. The handle of this heavy polished brass bell is made of two crouching lizards. The bell itself has two bands of intricate engraved designs, the bottom band with two dragons meeting on either side of a panel of Chinese letters. No interior markings. Stand: 19.5" high x 9.5" wide. Bell: 9" high. $75-100.

Valued for their history as well as their appearance, the little Mandarin hat button bells that hail from China are very popular among collectors. The Mandarins were high-ranking military or civil officials belonging to one of nine ranks. Governors and generals comprised the highest ranks, lieutenant governors and judges the middle ranks, and miscellaneous minor officers the lower ranks. To identify their rank, the Mandarins wore precious jewels of different colors on the tops of their hats, later replaced by semi-precious stones but still made of different colors. Towards the end of the Ch'ing Dynasty (c. 1900), these stones were made into the handles of little enameled bells, now known as Mandarin hat button bells. From one to nine (highest to lowest), the colors of the nine official ranks were as follows: ruby, coral, sapphire, lapis lazuli or blue opaque, crystal, moonstone or white opaque, plain gold, engraved gold, and silver.

Mandarin hat button bells from China. Left: Bright green repoussé enameled bell topped by an opaque yellow ornament representing the seventh rank. The five bats stand for the Five Great Blessings (happiness, health, virtue, peace and long life). 3.75" high. Right: Blue enameled Mandarin hat button bell of the fifth rank. 3.75" high. $125-225 each.

Left: Mandarin hat button bell with bats and conch design. Blue handle represents the third, or sapphire, rank. Right: Mandarin hat button bell with ruby ball, representing the first rank. $125-225 each.

Below: Pair of Chinese enameled bells with multi-color animals forming their handles. Left: Fish handle bell. 4.25" high. Right: Rooster handle bell. 4.75" high. $150-300 each.

Chinese cloisonné bell with "fishscale diaper pattern." Goldfish are favorites in Chinese decoration, symbolizing wealth and abundance. This bell has reinforced cloisonné work in the inside area where the clapper strikes. 7" high. $60-75.

Two sets of Chinese meditation bells or "healthy balls," made with sound plates inside that ring when gently shaken. Each of the two bells per set has a different tone, based on the principles of Yin and Yang. By stimulating the various acupuncture points on one's hand, the bells can—according to the folder that accompanies them—"adjust the nerve centre, benefit the brain, improve the faculty of memory, relieve the fatigue, drown the worries and help to get over mental fatigue." Blue set shown in and out of gift box. 2" diameter each. $10-20 per set.

Indian toe ring. Such rings were worn by Indian women as we wear rings on our fingers. The large ring slips over the toe and the rest of the decoration sits on top of the foot, tied with string or ribbon. Pellets inside rattle as the wearer walks. $35-50.

Pair of Indian ankle bracelets. Each is a two-piece bronze circle with nine crotals fastened to each half. Inside each semicircle are rectangular, rough depressions, possibly related to the casting procedures. $50-75 for pair.

The history of Indian "scrubber bells" is also quite interesting. While uncommon today, these small punchwork crotals, usually decorated with animals on their handles and containing small pellets inside, were formerly used in India for bathing or for scrubbing mud off the feet of women working in the fields. A raised pattern on the bottom helped with the cleaning process, although years of use often wore the underside smooth. These little scrubber bells were also used by blind masseurs in India to let potential customers know of their services; hence the bells are also known as masseurs' bells.

African mask with attached crotals and bell. This is a Dan tribal mask from the Ivory Coast of Africa, brought in by African traders. Around the carved wooden mask is a stuffed fabric edging with three small cowrie shells sewn at the top and sides. The bells are attached to the edging around the bottom half like a beard: four crotals on the left, three crotals and a bell on the right. The crotals are rather flat and primitive looking. 14.5" long. $200-250.

Set of three Indian scrubber (or masseurs') bells, brass. Handles of the two upright bells are most likely a water buffalo and a goat. The center bell is turned on its side to show the flat, scored base. $150-200 each.

Detail of the Sanskrit writing on the Nepalese bell.

Left: Large, very heavy Nepalese bell with Sanskrit writing around the base. The clapper extends slightly below the edge of this bell, which weighs approximately fifty pounds. 13" high. $200-250.

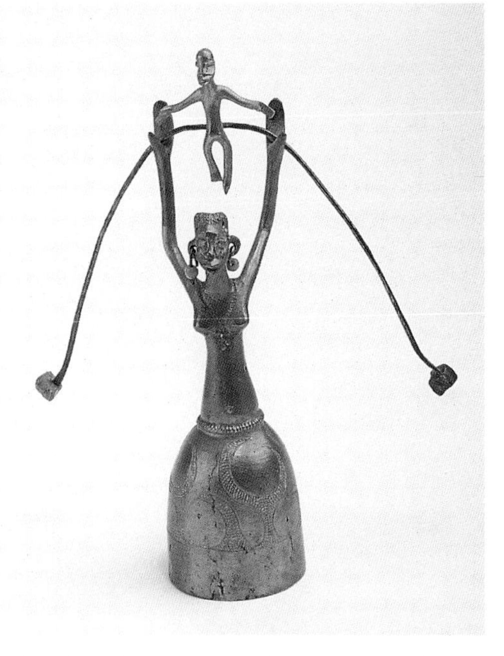

African figurine bell from the Ivory Coast, shown both front and side. The woman holds a balancing rod high above her head; a small child attached to the center of the rod swings back and forth freely in tandem with the rod, which is not attached. 5" high to top of head. $30-50.

Brass bell from Africa depicting a Cameroon tribal figure, with crocodiles decorating the base. The figure's elongated head is typical of African bells, in which body parts felt to be the most important are often exaggerated. Overall height: 5.5". $30-50.

Left: Globe-topped frog bell of heavy cast bronze from Cameroon, a country in western Africa. The handle is an openwork "globe," consisting of four big-eyed frogs leg to leg on the top half and four below, upside down. The top of the iron rod and ball clapper is split into two flanges which catch the top of the stem, leaving the whole clapper loose inside; it simply rests on the neck and is not attached. Overall height: 8.75". $150-175.

Right: Heavy, crudely cast brass bell also from Cameroon. The handle is a long flat "neck," slightly curved, ending in a face with fully defined features, a round beard, and a high rounded hat. The clapper is a simple long rod of iron bent into a hook at the top which goes through a metal loop. 8" high. $60-80.

Antelope horn bell, made by the Masai tribe in Kenya. The long wooden clapper is suspended by a leather thong which is in turn attached to a leather handle. 4.25" high. $25-35.

Wide base Egyptian bell inscribed with ancient hieroglyphic symbols. 5.5" high. $10-20.

Left: Although resembling an animal bell in shape, this primitive metal bell attached to a braided piece of rope is said to have been worn above the knee by African women, possibly as a decorative and musical accessory for dancing. $8-10.

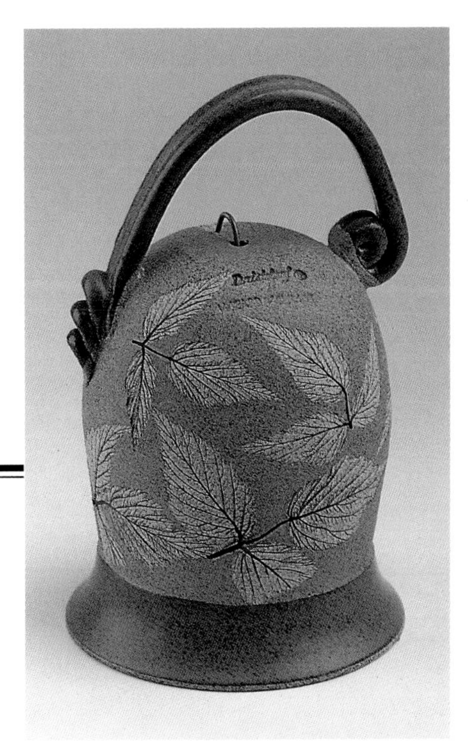

Partially glazed brown clay "Bristolleaf" bell, made at the Wizard of Clay Pottery in New York State. The distinctive feature of this bell is the imprint of raspberry leaves on the main body. At the Wizard of Clay Pottery, piles of fresh leaves are brought in each morning throughout the summer to create the Bristolleaf pieces. The leaves are pressed into the wet clay in artistic arrangements, then burn off when the clay is subsequently fired in the kiln. 5.5" high. $20-25.

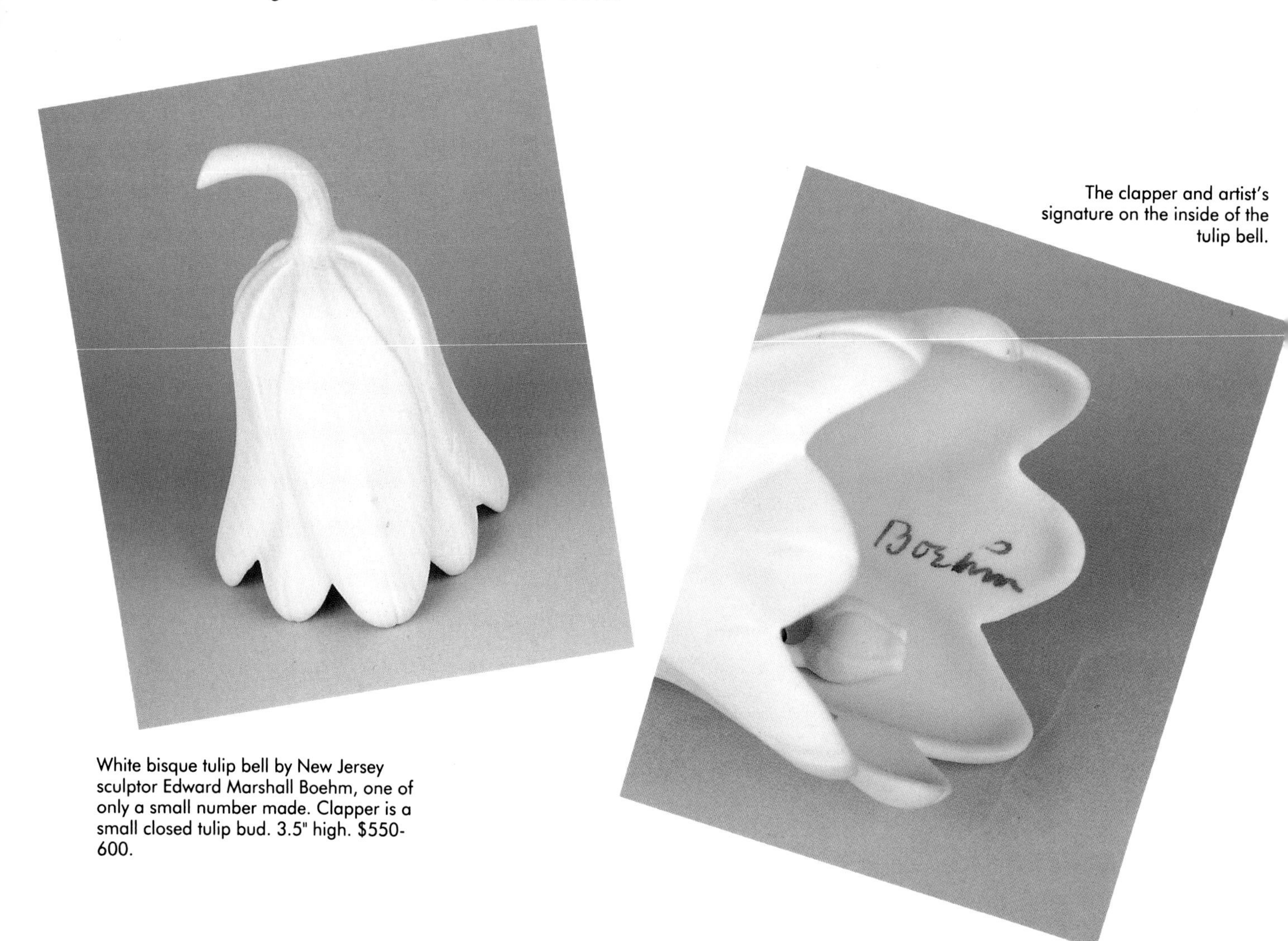

White bisque tulip bell by New Jersey sculptor Edward Marshall Boehm, one of only a small number made. Clapper is a small closed tulip bud. 3.5" high. $550-600.

The clapper and artist's signature on the inside of the tulip bell.

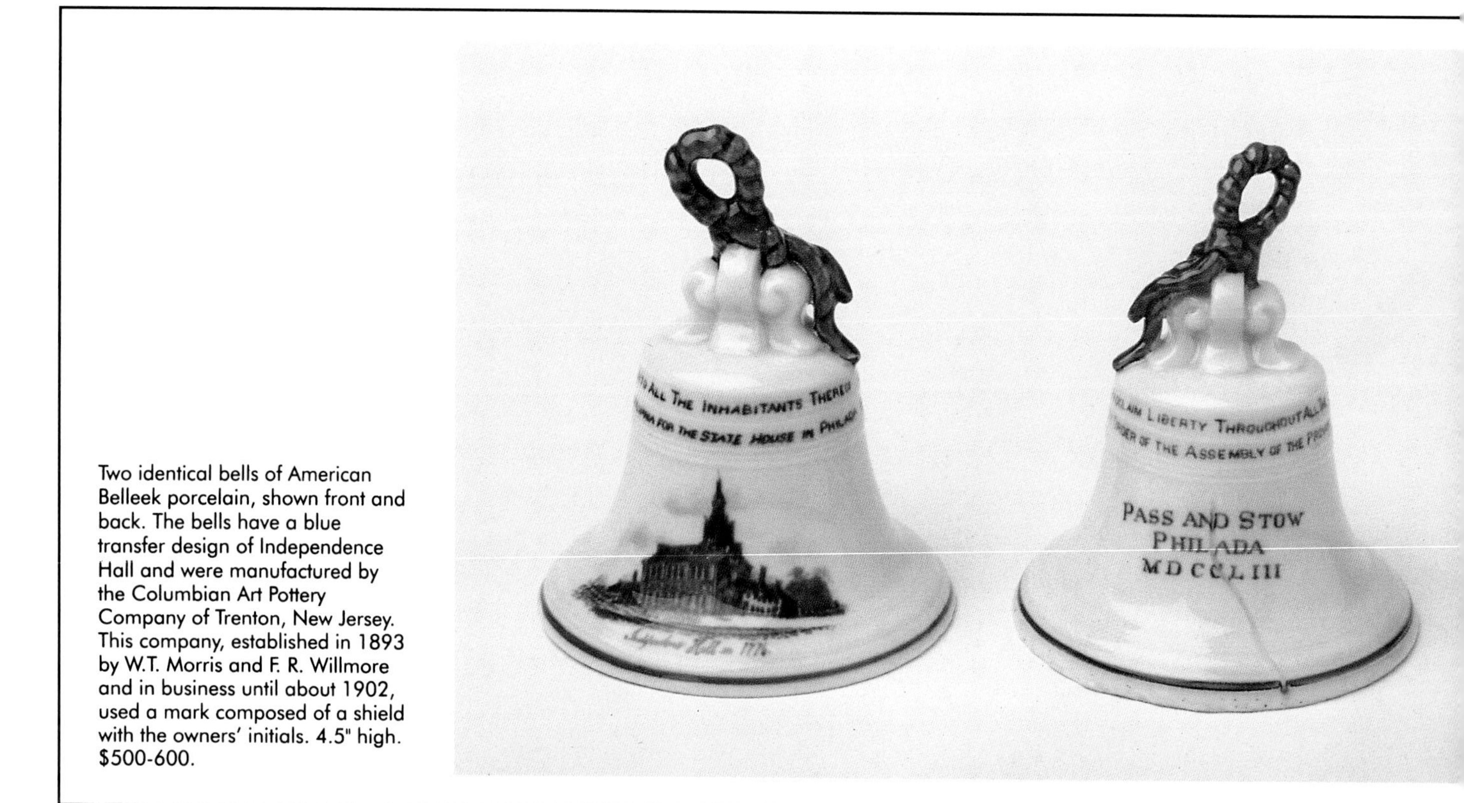

Two identical bells of American Belleek porcelain, shown front and back. The bells have a blue transfer design of Independence Hall and were manufactured by the Columbian Art Pottery Company of Trenton, New Jersey. This company, established in 1893 by W.T. Morris and F. R. Willmore and in business until about 1902, used a mark composed of a shield with the owners' initials. 4.5" high. $500-600.

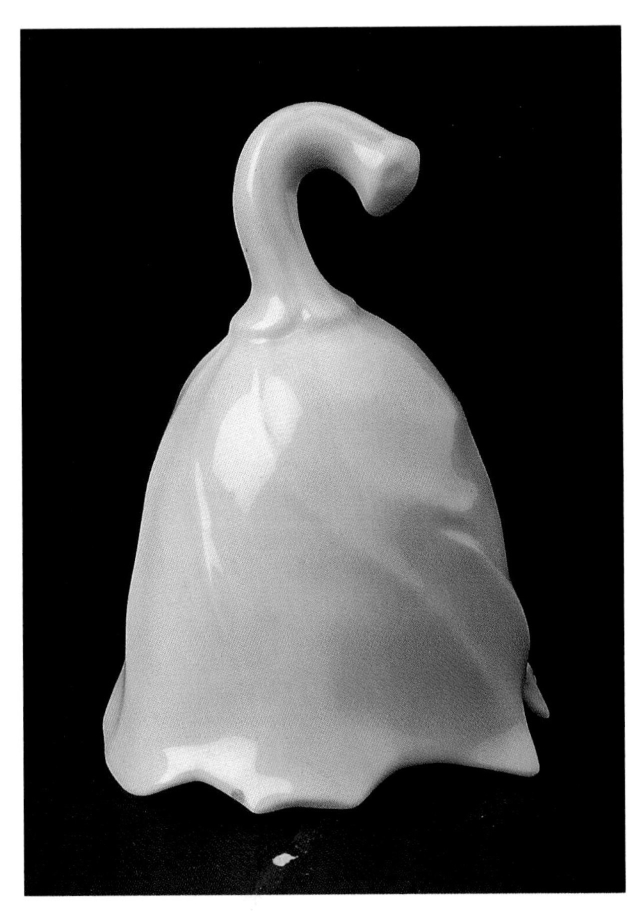

Another bell of American Belleek. Unmarked, but attributed to the Ceramic Art Company, founded in 1889 in Trenton, New Jersey. $400-500.

Cast in desert sand of the southwestern United States, this long hanging windbell was designed by Paolo Soleri, an Italian architect who came to Arizona in 1947 to study with Frank Lloyd Wright. Soleri is the originator and designer of Arcosanti, an experimental, energy-efficient community in Arizona based on Soleri's concept of "arcology," the integration of architecture and ecology. Proceeds from the sale of bells cast at Arcosanti help to fund the ongoing construction of this project. 27.5" long from clapper to end of hanging arm. $50-75.

Mark of the Columbian Art Pottery Company inside the American Belleek bell.

Ceramic bell purchased in Anchorage, Alaska, made by Carol McCrackin. Marbleized coloring of rust and cream with matching clapper. Painted inside: "Alaska Native Clay by Carol." 2.5" high. $15-20.

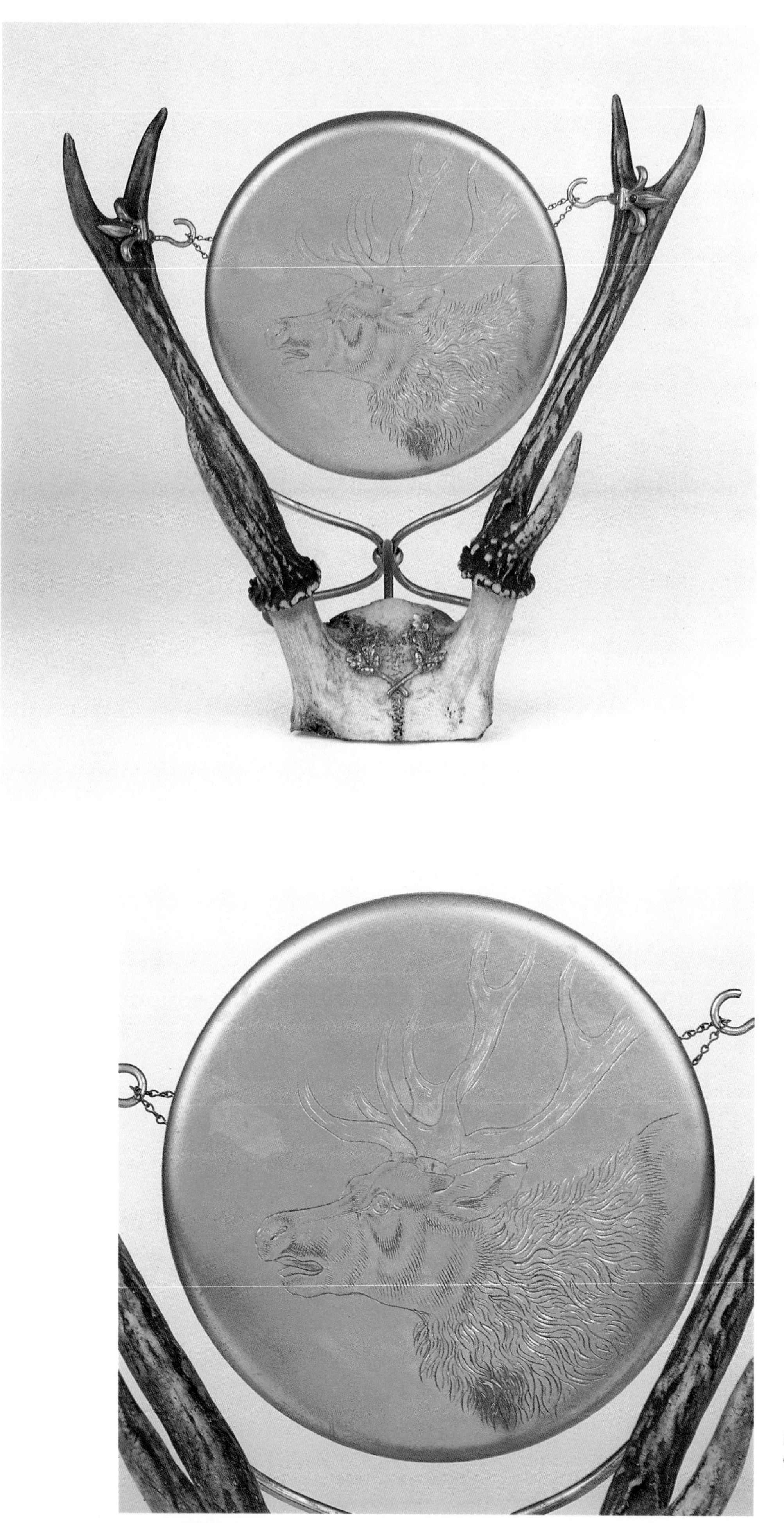

French Canadian brass gong with elk head engraving attached by fleur-de-lis holders to two deer antlers. Mallet used to sound the gong has deer hide hair inside to soften the tone. 6.5" diameter. $100-150.

Glazed pottery bell with whimsical painting of a sleeping dragon, by artist Louise Stockand of Vancouver Island, British Columbia. 5.625" high. $8-10.

Detail of the engraving on the deer antler gong.

Two copper bells from "down under," a kangaroo bell from Australia and a kiwi bell from New Zealand. Bell collectors live all over the world; the American Bell Association has chapters in both New Zealand and Australia. Both 3" high. $10-15 each.

Crude, unglazed clay bell from Mexico with a handle that consists of two animal heads back to back. One is obviously a red parrot with a big brown beak, the other looks like a calf with no horns. 4.5" high. $5-8.

No, this bell isn't *really* from Antarctica. But chances are, one that did come from that frigid continent would honor Antarctica's most prominent inhabitant—the humble but hardy penguin. 2" high. $3-5.

Chapter Four

Focus on Function

At one time, all of the bells in this chapter had a job to perform. While the relative importance of their functions varied, each used the sound it produced to accomplish a specific task. Their ringing might announce the start of school or church, might summon assistance or help, might keep track of a wandering cow or goat, might warn of unwelcome intruders, or might calm and distract a fussy child. In effect, the outward appearance of these bells could take any shape or form and they would still be effective, still able to carry out their respective functions. In reality, however, many of them are attractive as well as useful, valued for their exteriors in addition to their interiors. As you view each of these hard-working bells, you will find that their history and the story behind their original purpose lends them an unmistakable charisma and allure.

School, Farm, and Fire

Bells played an important role in everyday school and farm life prior to the advent of electricity and other technological advances. Although seemingly quaint and antiquated, old school and farm bells evoke a nostalgia and sentiment of high proportion in those who fondly remember "simpler times."

Large swinging school bells alerted children to the start of classes and gave those walking to school an idea of how quickly they needed to get there. Later, handbells were an essential part of every classroom. In contrast, students of today are regulated by the ear-splitting, insistent buzzers that signal the start and end of each class. Interestingly, however, the buzzers at most schools are still referred to as "the bell!"

1889 Birdville School bell. The iron piece to which the bell is bolted is called the yoke. 26" diameter. $350-425.

Anderson School bell from Butler County, Pennsylvania, with eagle on yoke, c. 1880s. Made of iron, diameter unknown. $300-400.

Assortment of classroom size school bells, with heights ranging from 3.5" to 9". Small sizes: $5-10 each. Large sizes: $45-65 each.

Russelton School bell made by the C.S. Bell Company of Hillsboro, Ohio, complete with iron wheel. 30" diameter. $500-600.

Close-up of the yoke on the Russelton School bell. The number in the center of the yoke gives the bell's size.

Parnassus Public School bell from 1892, made by the Fulton & Chaplin Manufacturing Company. 29" diameter. $2500-2900.

Silver painted farm dinner bell. 19.5" diameter. $150-250.

Right: No. 4 farm dinner bell. Iron, made prior to 1880. 20" diameter. $200-250.

Fire engine bell, unplated. 10" diameter. $700-800.

La France fire engine bell, 1933. 12" diameter. $700-800.

Bronze fire engine bell from New Castle, Pennsylvania, 1914. Made by Henry McShane & Company, Baltimore, Maryland. 16.5" diameter. $1400-1600.

La France fire engine bell with eagle perched on top, early 1900s. Unplated, diameter unknown. $700-900.

Transportation—By Land and by Sea

The clang of a locomotive bell is a familiar and pleasant sound that certainly enhances the excitement of embarking on a train trip or just watching as the train lumbers down the tracks. Old locomotive bells varied little in size and were used on the trains themselves as well as in the depots.

Ship's bells used in the United States date back to Revolutionary War days; unlike locomotive bells, their size (and sometimes their tone) was usually in direct proportion to the size of the ship that carried them. The bell was a highly significant part of sea life, tirelessly keeping busy sailors apprised of the daily schedule and even becoming a part of their lingo:

> Most important of all transportation bells have been the ship's bells. Even today, the ritual of ship's bells divides the time at sea into four-hour watches, when sailors take turns scanning the horizon and keeping an eye on the running of the ship. Before the coming of public address systems in ships, ships' bells were the way of announcing the change in watches. At every half hour, the bell is struck: once at noon, twice at 12:30, three times at 1 o'clock and so on until 4 when eight bells is struck. Then the bells begin all over again: one at 4:30, etc. The term "eight bells" is the one most commonly associated with sailors, for it is the time at sea when the watch changes. (Yolen 1974, 68)

Though infrequently found in the average collection, these hard-working transportation bells have a dignity that comes from a lifetime of service and are certainly worthy of our admiration and esteem.

Pennsylvania Railroad bell from 1878, non-original mounting. 11" diameter. $1100-1300.

Baltimore & Lake Erie railroad bell, No. 9627. 15.5" diameter. $1000-1200.

Pennsylvania Railroad bell, 1915. Value undetermined.

Regester & Webb brass railroad bell, from Baltimore, Maryland. 16" diameter. $1700-1900. *(See close-up below)*

Small gauge railroad bell from Punxsutawny, Pennsylvania. Value undetermined.

Close-up of Regester & Webb name.

Bronze ship's bell. Bell made by unknown foundry, yoke made by the C. S. Bell Company of Hillsboro, Ohio. 20" diameter. $900-1400.

New Jersey ship's bell, 18" diameter. $900-1200.

1833 ship's bell from the *Susquehanna*. The bell's inscription reads "Cast by J. Wilbank & Company." 16.5" diameter. $1600-2200.

Close-up of the inscription on the *Susquehanna* bell.

Pennsylvania canal lock bell cast by the A. Fulton Company of Pittsburgh, Pennsylvania in 1832. 12" diameter. $1000-1400.

Bells in Religion

Throughout history and throughout the world, ceremony and symbolism have played major roles in religious observance; bells have frequently been a part of both.

From sonorous Oriental temple gongs to resonant Christian church bells, the far-reaching sound of bells has solemnly convened worshippers for prayer and other religious services. As a part of the service itself, religious bells span a diverse and almost limitless range—examples shown here include the luminous and melodic Catholic altar bells and the ornately decorated Tibetan lamasery bells. Religious bells of still another form are those that pay homage to the respective gods or ecclesiastical figures of different faiths. These include the popular Evangelist bell venerating the four evangelists Matthew, Mark, Luke, and John, the Chinese enameled bells featuring figural handles of the Eight Immortals, and the Hindu Evolution bell honoring the wind god Vayu and the monkey nymph Anjana.

Henry McShane & Company bronze church bell from Baltimore, Maryland, c. 1886. Non-original stand, 19.5" diameter. $1200-1600.

Steel alloy Numderf Methodist Church bell weighing over six hundred pounds, manufactured in 1901 by the C.S. Bell Company. 40" diameter. $1200-1800.

Close-up of the Numderf Methodist Church bell yoke.

Crown top Spanish Mission bell from 1834. Value undetermined.

Above and left: Twentieth century version of an early Sanctus bell, which was placed upon the altar and used during the Mass. Christian symbolism is evident in the storks used to decorate this ornate bell, as legend describes a stork plucking feathers from its breast to make the manger soft for Jesus. Two fish, also symbolic in Christian art, form the bell's handle. Bell: 4.25" high. Underplate: 5" diameter. $150-175.

Bevin four branched altar bell with birds at the end of each branch. The four bells and birds signify the four evangelists. The bells are of polished brass, the branches and birds gilded with gold paint. The bells have triple clappers and produce a beautiful tone. 5" high. $50-75.

Two additional sets of altar bells, also with triple clappers. Such bells are tuned to a perfect major or minor chord. $35-50 each.

Triple clappers inside the altar bells.

Openwork metal bell on frame, mounted on a heavy marble base. This type of bell was originally used as a greeting bell outside of monasteries. 7" high. $100-125.

Evangelist bell, purchased in northern Scotland in 1959. Symbols related to the four evangelists are depicted in relief around the base. Above them is inscribed: "- S-MATHEVS:- JO HENNES - LVCAS: - S-MARCVS." Many variations of this bell exist, and it has been reproduced many times as an altar bell. 4.5" high. $50-60.

Another version of the Evangelist bell. Frequent differences include size, handle shape, and the amount of carving used for the evangelists' symbols. 5.25" high. $100-150.

A third Evangelist bell, larger and heavier than the others. 5.5" high. $100-150.

Hebrew spice box of sterling silver, used in traditional Jewish families as part of the ritual surrounding the end of the Sabbath day. The container is passed around to each family member and the scent of the sweet spices inside symbolize the final breath of the Sabbath. Encircling the central compartment which holds the spices are four vertical rods with flags at the top and small bells at the bottom. Stamped into the base is: "J. BERG STERLING." 8" high. $250-300.

Chinese enameled bells with figural handles depicting two of the Eight Immortals. The Immortals, six men and two women, were important figures from the ancient Chinese religion of Taoism and a popular theme in Chinese art. Each had a corresponding symbol, such as a crutch, fan, flower basket, or flute. The bell on the right has an ivory handle and also shows a Chinese symbol and five bats on the front. Left: 4" high. Right: 4.5" high. $150-300 each.

Another of the Eight Immortals graces the top of this brass and enamel bell. 5" high. $150-300.

Chinese bell with blue enameled flower design and figural handle depicting still another of the Eight Immortals. $150-300.

Heavy cast bronze Buddhist temple bell from Japan. The main feature of this bell is a very clearly molded bust of Buddha on both the front and back, showing the Buddha's hands in the traditional position with thumb and index fingers touching. The bell's clapper is a long cast rod and ball ending in a loop, obviously intended for a rope by which a worshipper could ring the bell at temple. This bell is from the Late Edo period, 1850-70 AD. 8.25" high. $200-250.

Drilbu lamasery bell from Nepal with eight mystic symbols encircling the base. Held in the left hand, this temple bell is rung to punctuate the readings of the Sutras. A dorje, shown next to the bell, is held in the right hand. Together they are the most important ritual objects of Tibetan Buddhism. Bell: 7.25" high. $50-75. Dorje: $35-50.

Crudely cast antiqued brass "bird monster" bell depicting Garuda, the mount of one of Hinduism's premier gods, Vishnu. In Hindu mythology, Garuda was a half man, half bird creature; he is usually shown as a man with the beak and wings of a bird. 5.5" high. $20-30.

At the top of this heavy bronze Hindu bell are Garuda and his consort, seated side by side with a "fan" of seven Naga (protective serpent demigods) above them. 9.25" high. $150-200.

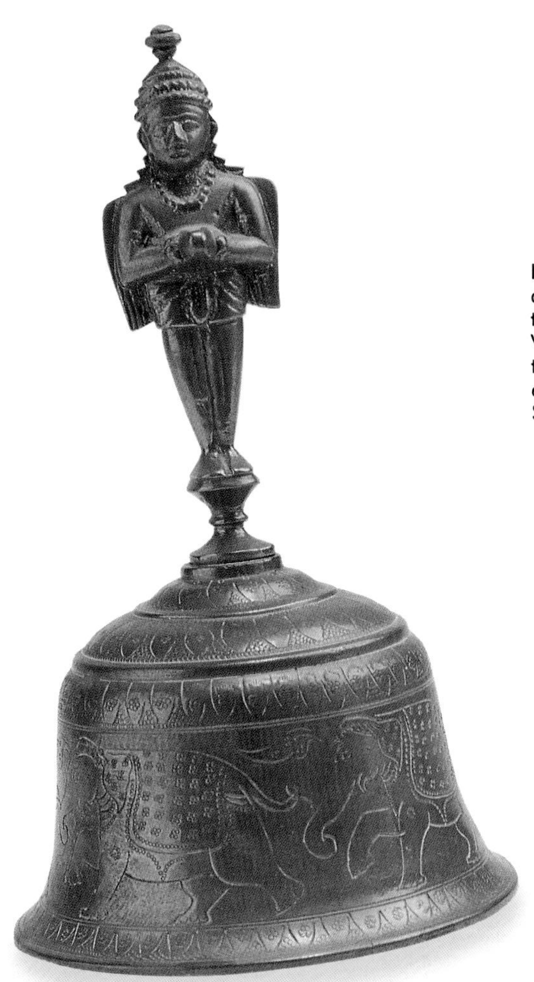

Hindu "Evolution" bell of cast bronze containing an incised pattern of marching elephants. The two-sided figure represents either the wind god Vayu and the monkey nymph Anjana (parents of the monkey god Hanuman), or Anjana before and after being turned into a monkey. 9" high. $50-75.

Hindu bell of heavy cast bronze depicting the monkey god Hanuman, son of the two figures from the previous bell. He is usually shown with his tail going up his back. 6.5" high. $75-100.

Reverse of the Hanuman monkey god bell.

Summoning, Signaling, and Securing

Here is an assortment of bells that use sound to summon assistance, get attention, or warn of intruders. Call bells, a somewhat generic term for bells used to attract the attention of another person or persons, have been used for hundreds—perhaps thousands—of years. Servant bells have been found in the tombs of early Egyptians, and no well-to-do house in Europe of the Middle Ages would be without one. Affluent families initially used small silver tea bells and elaborate figurines to ring for their servants, later the call bell evolved into a larger house bell that "hung like a church bell and was rung with a heavy cord that ended in a fancy tassel." (Yolen 1974, 64)

Distinctive French call bells, usually referred to as slave call bells, were first used on the boudoir tables of wealthy women to summon their personal maids. Later they became popular with the French who migrated to Louisiana during the time of slavery, hence the term "slave call bell." Mechanical models, they generally sit on a marble base and are lavishly decorated with mother-of-pearl shells and gold filigree borders.

Two examples of French slave call bells. Left: Hand painted mother of pearl shells around three sides, clusters of brass grapes, gold filigree brass circle on marble base, brass bird perching on top. Pressing the metal loop raises the tapper to sound the bell. 6" high. Right: French slave bell of similar design but without bird or painting. $250-400 each.

Another French slave call bell from the nineteenth century. Mother-of-pearl shells with star and half moon design around brass base. 4" high. $250-400.

Double-chiming call bell, patented in the 1860s by Ezra Cone of the Gong Bell Manufacturing Company, East Hampton, Connecticut. Also called a muffin bell or peddler's bell, it is rung by shaking in either direction. The same bell can be found in various sizes; this one is 10.5" long. $75-100.

Hotels, restaurants, and stores still use tap bells on occasion, so customers can announce their arrival and be efficiently served. While serving a useful purpose, these bells of today are rarely as pleasing to the eye as those of yesteryear. The same might be said of doorbells, which now tend to be pragmatic and unobtrusive at best but were often handsomely decorated and miniature works of art in the past.

Silver tap bell. The ornate, "bud" shaped clapper is activated by depressing the tap mechanism on top of the bell. Engraved around the base of the plunger just above the canopy is: "PAT^S AUG. 25. 1863 & APR. 8. 56." This type of bell was often used in hotels, stores, and other places where customers or guests needed to make their presence known. 4.5" high, 3" diameter base. $25-50.

Left: Pewter and brass tap bell. When purchased, the stone on the top of this bell was missing. The gray stone seen here was purchased separately and attached to the top to match the base. 4" high. $50-75.

Silver-plated Victorian hotel desk set, shown from two views. Combination tap bell, inkwell, and pen holder. Below the tap on the bell is incised: "PAT.D' AUG. 25, 1863, AUG. 8, 1856." Overall height: 11". Bell: 5.25" high. $150-200.

Next to mastering the art of staying upright, bicycle riders of all ages find great pleasure in announcing their presence and signaling their turns with the ever-present bicycle bell. Those shown here will surely bring back fond memories for many two-wheel aficionados and prove a sharp contrast to the sleekly designed, high tech models often used today.

Finally, bells can also serve as alarm mechanisms, warning of doors being opened inappropriately and providing a measure of security for those worried about potential intruders. The tone of these bells is far more discordant than melodious, but that's the idea!

Twist doorbell, rung by turning a handle on the outside of the door. The handle emerges from an ornate circular plate with much scrollwork and a beaded edge. 2" high, 3.75" diameter. $50-65.

Metal bicycle bell with thumb trigger. 3.5" across trigger and bell. $20-25.

Left: Old Bevin bicycle bell. Circular frame with handlebar mounting clamp and thumb depressor to ring the bell. Incised into one side of the thumb depressor is "BEVIN PAT. #D109032" and on the other side, "MADE IN USA." Initial design c.1937. Right: Snap type bell, most likely made for a bicycle. The metal disc is struck by a snapping, spring mounted rod with a ball on the end. The bell is attached to a circular clamp and screw of cast iron, indicating that it was made to be mounted on a round bar of some kind. This mounting unit, combined with the hand activated snapper, suggests this bell's original use as a bicycle bell. $30-50 each.

Unique display of bicycle bells mounted on an old handlebar. The bells include those activated by thumb depressor, twirl (far right), and push button.

Burglar alarm bells made by the Surety Alarm Company, c. late nineteenth century. The bell is wound by twisting the top, then used by placing the angled edge under a door. If the door is opened the bell inside will sound loudly to scare off intruders. 4.5" and 3.75" long. $40-60.

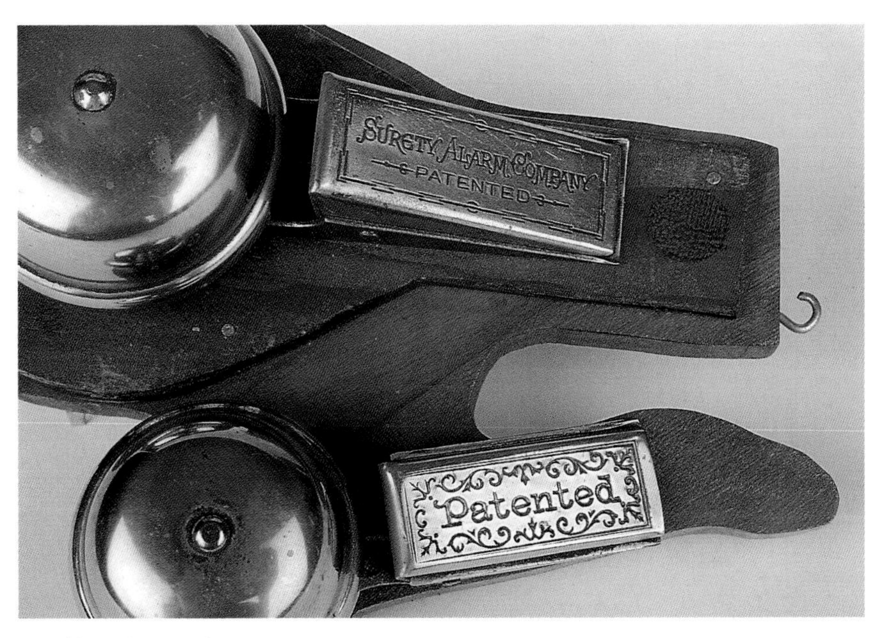

Detail from the top of the two burglar alarm bells.

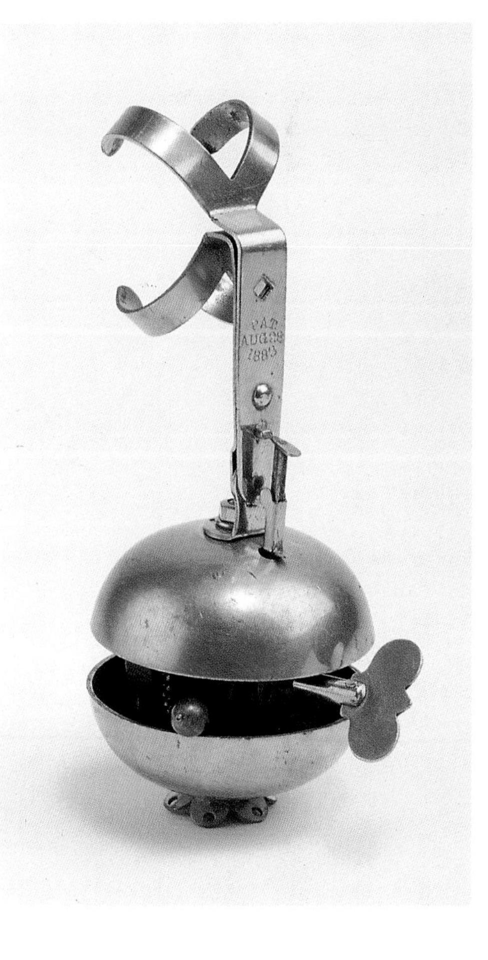

Swinging door alarm bell, comprised of two metal hemispheres separated by a wind-up ringing mechanism. The bell can sit on its six petaled "foot" but is really meant to be hung from the swiveling metal stem. To make the bell work, a teardrop shaped guard piece is first raised up to rest on a protuding pin from the vertical stem. This allows the bell's spring to be key-wound without ringing and also sets the mechanism. Next, the bell is hung over a doorknob. Should the doorknob be turned, the guard piece is thrown off the pin and the bell rings with a loud alarm. Incised on stem: "PAT. AUG. 28 1883." $50-75.

Animal Bells

Almost everyone is familiar with the cowbell, and one of these traditionally shaped old bells is often the first of many acquired by prospective collectors. Yet animals of all kinds and from all countries have often been "belled," as a means of keeping track of them, warding off predators, or simply providing them with an attractive adornment.

The earliest animals to wear bells may have been sheep from the flocks of ancient shepherds. Initially all sheep in the flock were belled, later only the lead sheep was given this distinction. Since "wether" was the term for a neutered sheep, this lead sheep was sometimes known as the "bell-wether." Today, this word is still used to denote a leader, albeit one of a foolish, i.e., "sheeplike" crowd! Other animals that have worn bells through the years include horses, donkeys, goats, water buffalo, oxen, elephants, reindeer, birds, camels, dogs, cats, geese, and pigs. (Yolen 1974, 42)

Other than cowbells, perhaps the most widely recognized animal bells are the popular straps of small jingly sleigh bells. In reality, however, sleigh bells were only one of a great variety of bells that have been used on horses. Lois Springer describes the many purposes attributed to horse bells through the years:

> Each culture evolved its own style of belled [horse] trappings, but the reasons for their use, real or imagined, remained fairly constant: they frightened off any evil spirits lurking near the horse's path; on narrow, winding trails they warned of his approach; and they instilled confidence, even courage, in the animal himself, especially in warfare. On special occasions they also added pomp and gaiety to ceremonial processions. (Springer 1972, 147)

Brass cowbell marked "1878 CHIANTEL SAIGNELEGIER FONDEUR." Collectors who purchase this bell often think they have found one dating from the nineteenth century. In reality, however, the 1878 date on the bell does not mean it was cast in that year. The foundry which made these bells cast the *first* one in 1878, but continued using the same mold bearing that date well into the next century. 3.75" high. $25-35.

Bevin Brothers cowbell with original label. 4.5" high. $40-50.

Two types of cattle bells. On the left is a Holstein Bell No. 7 with a cow pictured on the label. The Australian "Condamine" bell on the right has an interesting history that is described in Lois Springer's *Collector's Book of Bells*. Springer notes that such bells were originally made by S.W. Jones in an area of Australia near the Condamine River more than a century ago, and were labeled "Bull Frog" for their hoarse, frog-like tone. This Condamine bell was made by Alf Ormand, who succeeded Jones in 1912 and was the only one of his assistants to successfully replicate the bull frog sound. Holstein bell: 3.25" high. $20-30. Condamine Bull Frog bell: 2.75" high. $25-35.

Diminutive cowbell with engraving of farmer and cow. The front reads "SIMOND CHAMONIX." 1.5" high. $20-25.

Four cow or sheep bells. From left: French cowbell, 3" high; Swiss cowbell, 5" high; Basque sheep bell, 4" high; Greek animal bell, 1.75" high. $15-25 each.

Close-up of the Swiss cowbell.

Old and battered sheep bell with four blue balls decorating chain. 3.25" high. $10-15.

Some of these ceremonial appointments for horses included saddle chimes and swingers, several examples of which are shown here. The swingers, also known as fliers, consisted of a tall stiff plume decorated with one or more brass bells. They could be attached to the horse's saddle or used on the horse's head as part of the harness. Either way, they made for a most noble and stately appearance, and were, in fact, primarily used by those of noble or wealthy status.

Horse swinger with two bells, originally one of a pair. This one has the original three layer brush top with center layer of darker color. It is rare to find these intact. Overall height: 9.5". $50-75.

Saddle chime, used for decorative purposes on horses. The two bells on the side have three clappers each and are fastened to the harp by brads. The top bell consists of two cups on a screw stem with a spider of six external clappers on top. Overall dimensions: 9" high, 6.75" wide. $50-75.

Another set of saddle chimes. $50-75.

Saddle chimes for donkeys, from the Italian island of Sicily. Used by placing over the donkey's back, with the reins threaded through the two large circles. Supposedly, the number of bells on such pieces signified the wealth of the donkey's owner. $450 and up.

Sleigh bells, the familiar crotal style bells that come in different shapes and sizes, were originally developed to help avoid "traffic collisions" by multiple sleighs traveling noiselessly along slick, snowy roads. They were made as early as colonial times, when the two halves were cast separately and soldered together after a pellet clapper was placed inside. In the early nineteenth century, a method of casting the bells all in one piece was developed by William Barton of East Hampton, Connecticut. Barton's invention, in which the pellet was encased in a core of sand that was later sifted out, led to his reputation as the most significant figure in sleigh bell history and to a long legacy of sleigh bell manufacturing in the Connecticut Valley. Barton's son and grandson, Hiram Barton and William E. Barton, succeeded him in the sleigh bell industry and many others learned and apprenticed under his tutelage as well. Around the middle of the nineteenth century a technique for "stamping" sleigh bells out of brass was developed, proving much less expensive than the earlier casting procedures.

Set of twenty-one nineteenth century sleigh bells riveted to a leather strap. Each bell marked with patent information as follows: "Oct 24 68 May 14 78." $125-150.

Set of twenty-four acorn shaped sleigh bells, most likely manufactured by Bevin Brothers. Restored and riveted onto a new leather strap. $80-100.

This set of eight sleigh bells would have been used around the horse's chest instead of being wrapped around his body. They are attached to the leather strap by a method called the "Cotter key" and were manufactured by Hiram Barton. $100-150.

Detail of the Cotter key attachment holding the bells on the strap.

Sleigh bells were made in twenty sizes, ranging from approximately 4" in diameter to less than 1" in diameter, in several different shapes, and with single, double, or triple "throats" (slits) across the bottom. The bells were either riveted to leather straps or toggled to the strap using a variety of methods. Fewer of the large sizes were made; as a result they are more difficult to find. Sets with graduated sizes are the most highly prized among collectors.

Graduated set of twenty-three sleigh bells with sizes ranging from 5 to 13, attached by Cotter key to a double leather strap. The double leather was used to protect the horse from the sleigh bells. $350-400.

Sleigh bell trio representing three generations of sleigh bell manufacturers. From left: William Barton, Hiram Baron, William E. Barton. $25-35.

Double strap of pony bells, with sizes ranging from 0 to 3. $150-175.

Set of thirty raspberry sleigh bells with original riveting. $90-125.

Above: Shaft chimes or thill bells. Each bell has three clappers so the bell would ring with the slightest movement of the sleigh to warn people of its approach on snowy roads. They came in pairs and were used on the two shafts that ran along each side of the horse. $35-50.

Interior of the shaft chimes, showing the three clappers inside each bell.

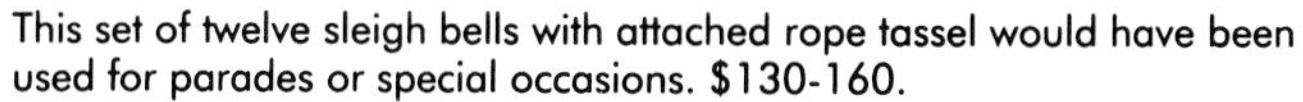

This set of twelve sleigh bells with attached rope tassel would have been used for parades or special occasions. $130-160.

Goat bell on a wooden yoke. Worn around the goat's neck with a leather strap holding the bell. 12" long. $20-25.

Primitive animal bell from the Masai tribe in Kenya, probably used for goats. Metal bell with handle made of some kind of plant fiber. $10-15.

Water buffalo bell from the Philippines. Crotal style bell with four triangles and two eye shape openings cut into the body of the bell. 3.75" high. $15-20.

Helmet shaped water buffalo bell, Javanese. The bell is made of hammered brass, the clapper is a tubular length of wood (possibly a branch) suspended through a small hole below the handle by a piece of cord. Very blurred marking above the lip on one side reads "JAVA." 4.25" high. $15-20.

Clapper from the Javanese water buffalo bell.

Wooden oxen bell from southeast Asia. Clappers are two hinged wooden balls on either side of the center bell." 4.5" high. $10-15.

Bells may seem a somewhat incongruous adornment for such ponderous animals as elephants, yet elephant bells are much favored among collectors and have a multitude of uses. Besides identifying specific elephants with specific tones, the bells help provide working elephants with a rhythmic cadence as they walk, alert others that elephants are coming through the area, and serve as "joyous embroidery decorating the wise and happy elephant." (Shayt 1997, 24)

Claw-type elephant bell. 2.25" diameter. $3-5.

Cast bronze elephant bell from Thailand, believed to be more than 150 years old. 2.75" diameter. $25-30.

Animal bell from Thailand, most likely an elephant bell. The bell is mounted on a wooden stand painted black. Overall height: 6". Bell: 4.5" high. $35-65.

Camel anklet bell, made in India. Crescent-shaped bell with grotesque faces at each end and inscription on the back side. Similar type bells were used on camels that were trained to dance; the bells may also have been worn to ward off snakes. $25-40.

Set of old Persian camel chimes consisting of four bells one within the other. 11" long. $40-50.

Detail of the inscription on the camel anklet bell above.

A colorful story lies behind this somewhat dignified looking bell known as the US Cavalry bell. In the late 1850s, camels were imported by the US Army under the direction of Secretary of State Jefferson Davis and used to transport supplies to army posts in the southwest. Although this experimental use of camels for a mounted cavalry unit failed after only a few years, bells made by the Starr Brothers Bell Co. of Connecticut that were supposedly used on these "ships of the desert" have become quite collectible. Whether this bell ever served as the actual camel bell has not been fully authenticated, however many replicas identifying it as such were successfully manufactured and sold by a bell foundry in California. 4.25" high. $15-30.

Reindeer bell made of heavy cast bronze with a square base. On the front is the head of a reindeer in relief. In Scandinavia, reindeer herds are belled to help locate the animals in snowstorms. 2.75" high. $25-30.

Eagle bells, hand made of silver and bronze with each piece soldered to the other. Used for hunting, their purpose is help locate the bird or quarry since hunting birds do not retrieve their catch. Such bells are designed in different shapes and sizes for other birds of prey and are made by Peter Asborno of Denver, Colorado. Two crotals, each 1.25" high. $15-25.

Wartime Bells

Bells have played many roles in the course or war or battles, and several with interesting stories and from different eras are included in this section. A popular bell for collectors is the British "ARP," or Air Raid Patrol bell. A simple handbell in style, its sharp clanging was intended to warn citizens of the need to seek shelter during the air raids of World War II. Most valued are those cast by the Gillett and Johnston foundry; these bear the identifying initials "G & J."

Two of the bells shown here were not actually used during wars, but are included because of their connection to peacetime efforts directly resulting from wars. First is the V for Victory bell, known more for its patriotic and historic value than for its somewhat unassuming appearance. Designed by the British Royal Air Force (R.A.F.), the bell is made of unknown metal, gray in composition. Depicted in relief around the base are profiles of the three Allied leaders from World War II: Joseph Stalin, Franklin Roosevelt, and Winston Churchill. An inscription carried around three sides of the lip identifies the origin of the gray metal: "CAST WITH METAL FROM GERMAN / AIRCRAFT SHOT DOWN OVER BRITAIN / 1939-45 R.A.F. BENEVOLENT FUND."

Revolutionary War bell from 1776, painted red, no original mountings. Approximately 12" diameter. $500-900.

Used during World War I, the purpose of this unusual looking gong is not immediately apparent. The mallet, shown across the top of the gong, pulls out of one end and was banged loudly between the prongs of the inverted metal "U" to alert soldiers of the need to don gas masks. The gong could be worn around the neck and the mallet held in place with a pin. 19" long with mallet in place. $20-25.

Cargo light with bell, also known as a battlefield chaplain's lantern. Heavy brass lantern with glass panels, two vertical bars on each side, and a Saignelegier/Chiantel Fondeur bell attached to base. Molded plate on top reads: "CARGO LIGHT, No. 3954, GREAT BRITAIN, 1939." A replica of the kind used in nineteenth century wars, this cargo light includes a crucifix on the front for administering last rights, a candle for illumination, and a bell to warn the enemy that the person carrying the light was a chaplain. 16" high, 6.5" square. $150-200.

Right: Air Raid Patrol bell. Heavy wooden handled bell, obviously well used, with varnish on the handle chipped and worn off in many places. "ARP" is stamped on the lip. The bell is also stamped "G & J 1939" for the Gillett and Johnson foundry which cast it. 9.75" high. $75-100.

Below: Close-up showing the letters "ARP" on the lip of the bell.

V for Victory Bell, also known as the R.A.F or Yalta bell. Encircling the base are profiles of the three Allied leaders from World War II, who met at the Yalta Conference in February, 1945 to discuss the defeat of Germany. 5.75" high. $25-50.

This second peacetime bell illustrates how the horrors of wars past can sometimes be transformed into a message of hope and new development. In Cambodia, an organization known as the Church World Service (CWS) is working towards that end through a partnership effort they call "Shells into Bells." As they work or play in the fields, Cambodian children gather scrap metal left over from the Khmer Rouge reign of terror in the 1970s. This scrap metal—spent shell casings, land mines, and even unexploded artillery shells and bombs—helps increase the family's income by being sold to the village foundries. There it is melted down and recast into tiny bells, bells that are used on cows, oxen, and water buffalo but that are also symbolic of the Cambodian people's commitment to a better way of life.

"Shells into Bells" crotal type bell made in Cambodia of wartime scrap metal gathered by children working in the fields. $5-10.

Rattles and Toys

Captivating and innately cheerful, bells have been used for centuries to amuse and delight children. Who among us has not smiled as an infant's eyes light up with delight at the sound of a tinkly bell? Today's children are no different from those of long ago in their universal enjoyment of bells.

Lois Springer writes in *The Collector's Book of Bells* that the sixteenth through the eighteenth centuries constituted what can be called the "golden age of baby bells." (Springer 1972, 190) While many different kinds of gold and silver bells were produced during that period, the most prized examples are those known as coral-and-bells, also referred to as whistle bells. Cylindrical in shape, these rather ornate looking devices were multi-functional: the whistle at the top could be used by a nanny or nursemaid to gain a child's attention, small bells around the body amused or perhaps distracted the child from crying, and the long, narrow bottom of coral, ivory, or a similarly hard substance helped to soothe a teething child's sore gums.

Various styles of sterling silver baby rattles with bells. Up to $100 each.

Teething ring with bell; metal baby rattle with wooden handle. $10-15 each.

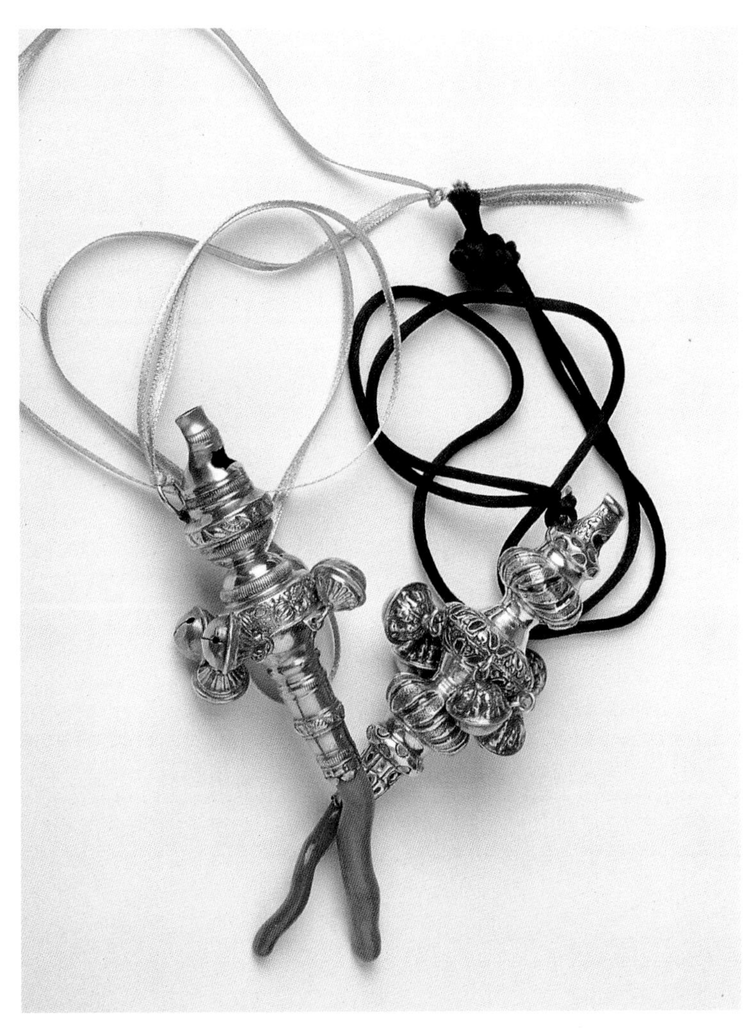

Two whistle bells, also known as coral-and-bells. The sterling silver example is English, while the gold was made in 1819 by Samuel Pemberton and given to a child in North Carolina as a gift. The 1819 date and maker's initials are inscribed at the top of the gold whistle bell. Both 5" long. Sterling silver: $400-600. Gold: $700-$1000.

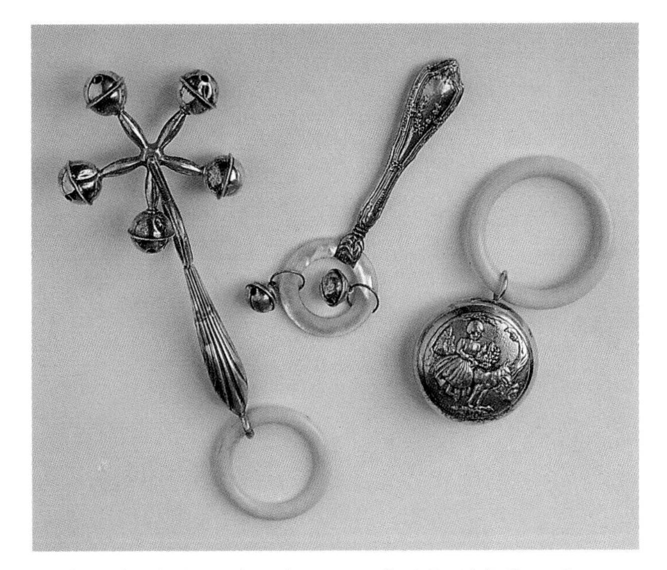

Sterling silver baby rattles. The one on the left with bells on five spokes is Italian, the one on the right with an illustration of Little Red Riding Hood is Dutch. Up to $100 each.

For older children, metal toys known as bell ringers came into vogue during the late nineteenth century and were widely manufactured from about 1880 to 1920. Using a combination of wheels and bells, these toys were favored by children for their ability to generate a ringing bell sound when the toy was pushed or pulled along the floor. Animated bell ringers were those in which the movement of a figure, either human or animal, was included in the process of ringing the bell. (Springer 1972, 196)

Metal toy with rider pulling a wheeled bell behind his galloping steed. 7" long. $200-225.

Bell ringer toys, c. late nineteenth century. Left: 3" high, 3.5" wide. $50-75. Right: 4" high, 4" wide. $100-125.

Centennial bell toy from 1876. During the American Revolution, the Liberty Bell was taken from Philadelphia to Allentown, Pennsylvania and hidden in the basement of a church for safekeeping while the British occupied Philadelphia. This metal toy, complete with American flag, portrays the Liberty Bell being transported via farm wagon on its historic journey in 1777. 8" long, 5.75" high. $400 and up.

Chapter Five

In Celebration of Shape and Style

This chapter presents an array of bells illustrating the wonderful variety of designs, textures, colors, and decorations that make each bell so unique and pleasing. While sound certainly enhances the interest and charm of all bells, those presented here are valued primarily for their appearance, whether beautiful, comical, historical, or representative of a specific person, place, animal—even another bell. Here you will find bells purchased as mementos of trips or special occasions, bells that replicate many of the world's largest bells, bells honoring famous people, bells with animal themes, bells used as holiday decorations, bell jewelry, and more. Some are widely available in gift shops and stores, others rare, uncommon, or even one-of-a-kind. Each is interesting and desirable in its own right, a welcome addition to the new or advanced collection.

Souvenir Bells

Many collectors find the purchase of an inexpensive but attractive bell to remind them of a memorable trip or a favorite place visited has been the spark that ignites a long term fascination with bells. Easily found in most souvenir and gift shops, bells take up little room on the way home and serve as fond mementos of special vacations. They also make handy gifts for both current and prospective bell lovers!

The original purchase price for most souvenir bells ranges from $5-10. Values for these bells are not listed individually, as they have essentially no resale or market value. Nonetheless, the lack of market value for souvenir bells is usually compensated for by the high degree of "sentimental value" they afford to their owners.

A trio of bells brought home from trips around the country. From left: Massachusetts, 5.75" high; North Carolina, 4.25" high; Wildwood, New Jersey, 5" high.

Souvenir bell from Louisville, Kentucky, depicting the famous Churchill Downs Racetrack. Cream and blue stoneware, heart-shaped handle. Made by Louisville Stoneware, Inc. 6" high.

Florida flamingoes grace these two souvenir bells from the Sunshine State. Both 4.75" high.

Another two bells brought home from Florida. Both 5" high.

Souvenir bell from Birmingham, Alabama. 5.5" high.

Souvenir bell from Columbus, Ohio. 4" high.

A moose is featured on both of these souvenir bells from the state of Maine. Left: 4.25" high. Right: 3" high.

Two versions of "The Big Apple," souvenirs from New York City. Left: 3" high. Right: 4.25" high.

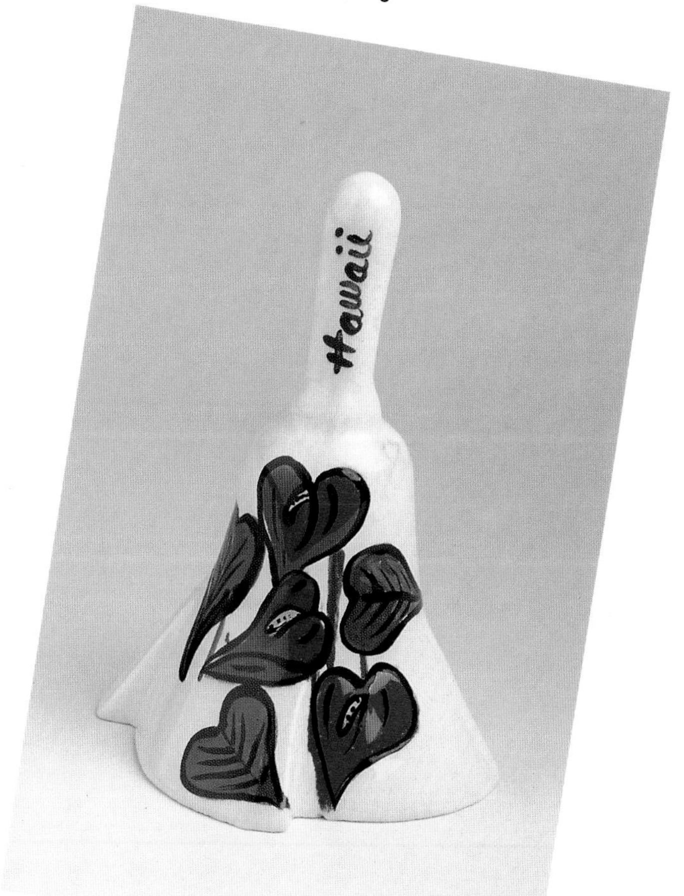

Souvenir bell from Hawaii. White ceramic with painted red anthuriums. 4" high.

Hand painted porcelain bell with wooden handle from Minnesota. Loons on lake design, artist's initials "JH" on outside.

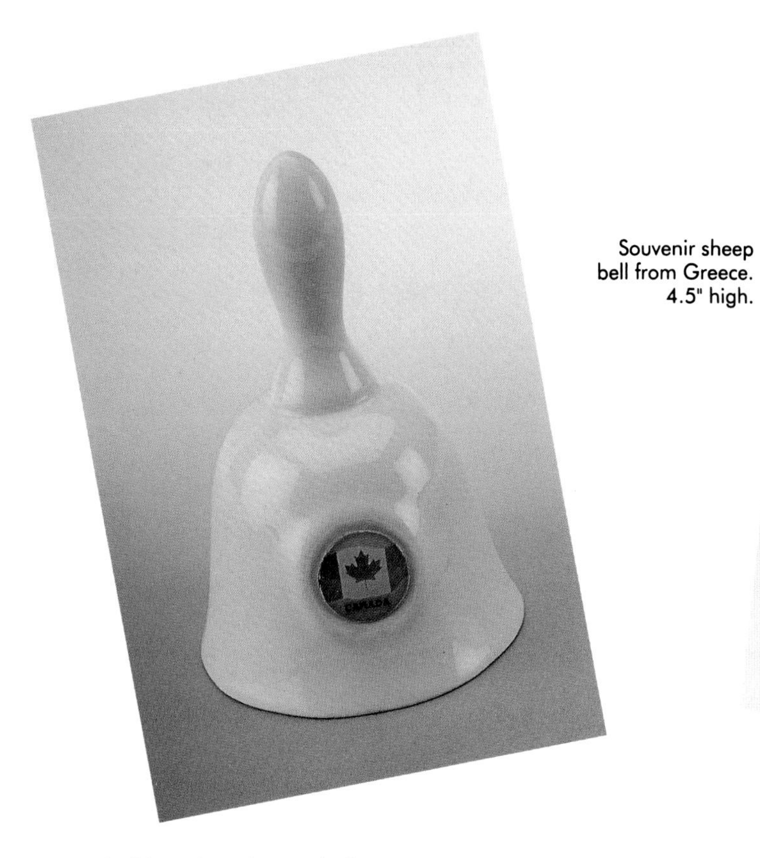

Souvenir bell from Canada. 3.5" high.

Souvenir sheep bell from Greece. 4.5" high.

Memories of a trip to Madrid belong to the owner of this brightly colored souvenir bell from Spain. 2.75" high.

This bell came from the balmy Bahamas. 5.25" high.

Commemorative Bells

Bells that honor or memorialize an event, rather than a place, can be aptly categorized as commemorative bells. Events of national, historic, or political significance—as well as their subsequent anniversaries—lend themselves well to depiction on bells. Traditional open-mouthed bells seem particularly befitting commemoratives for royal events in Britain, so steeped in pageantry and ceremony, and three examples are shown here. In America, the popular expositions and world's fairs from the last century all produced commemorative bells of fine workmanship, many now admirable additions to a collection.

Commemorative bells are certainly not limited to events of national importance. Modern wedding guests often take home small bells inscribed with the names of the bride and groom and the date the marriage ceremony took place. Conventions and meetings may use bells as favors or keepsakes. In reality, any occasion that holds importance for those in attendance can be remembered and virtually immortalized through its later depiction on a commemorative bell.

British bell commemorating the fortieth anniversary of the Coronation of H.M. Queen Elizabeth II, 1953-1993. 4.5" high. $25-35.

The reverse of the coronation bell lists the names and dates of England's former kings and queens.

Front and back of Royal Crown Derby bell commemorating the Investiture of H.R.H. The Prince of Wales in 1969. Marked inside number 200 from a Limited Edition of 500. 8" high. $150-200.

Commemorative bell celebrating the 500th Anniversary of Heidelberg University in 1886. 6" high. $125-175.

Three bells commemorating twentieth century British marriages. From left: Princess Anne wedding bell of Aynsley English bone china. The back reads: "To commemorate the marriage of H.R.H. The Princess Anne to Capt. Mark Phillips, Westminster Abbey, November 14th, 1973." 5" high. $25-30; Prince Charles and Lady Diana wedding bell of white porcelain by Caverswell. 3" high. $15-20; Prince Andrew and Sarah Ferguson wedding commemorative of glazed porcelain. 3" high. $15-20.

Most likely a bell issued in conjunction with a convention or meeting, this maroon, fez-shaped bell from the Zuhrah Shrine includes the characteristic black cord and tassel. 3" high. $10-15.

These five, finely detailed glass bells are all commemoratives from the 1893 Columbian Exposition held in Chicago, Illinois. Flower decoration bell: 5" high. All others: 4.5" high. $50-75 each.

Fenton Bicentennial bells, commemorating the 1776-1976 American Bicentennial, in iridescent blue and red. The bell body is decorated with four medallions in relief of George Washington, Thomas Jefferson, John Adams, and Benjamin Franklin. An eagle on top of the handle has its wings partially unfurled. Each bell has the Fenton trademark molded inside at the top and an "Authentic Fenton Handmade" sticker inside. Both 6.5" high. $50-75 each.

"Let Freedom Ring" bell. Hand painted porcelain eagle mounted on a lead crystal base shaped like the Capitol dome. The clapper is a minted gold "coin" with heads of eagles on both sides. This bell was made by the Franklin Mint to celebrate the 200th Anniversary of the Congress of the Unites States; the artist is Ronald Van Ruyckevelt. $200-215.

Two contemporary bells with a political theme. The bell on the right illustrates "The Choice" from 1980, with portraits of Carter, Reagan, and Anderson on the front. The Inaugural Issue bell on the left commemorates the results of that choice with a portrait of Ronald Reagan on the back. Both are Limited Editions of 5000. 3.25" high. $15-25 each.

This commemorative bell represents an important piece of bell collecting history! The tiny 1.25" ceramic bell is marked "1948" on one side and "NBCC" on the other. NBCC stands for National Bell Collector's Club, which was the original name for the organization founded in the early 1940s for bell fanciers. At its 1948 convention the National Bell Collector's Club changed its name to the American Bell Association, still in use today. This bell has a small slot in the handle indicating that it may have been used as a placecard holder at the convention. $10-15.

Replicas of Famous Bells

In recognition of the history and beauty behind many of the world's largest or most well-known bells, far smaller replicas with realistic detailing are popular additions to a collection of "household size" bells. Bells from both Europe and America have been honored through such reproduction on a smaller scale.

Among massive European bells, the Tsar Kolokol and St. Peter's bells are two of the most frequently seen in replica form. Also known as the Great Bell of Moscow, the two hundred ton Tsar Kolokol stands in the Grand Square of the Kremlin in Moscow. It rests over the spot where it was cast in 1734, but the great bell has never actually rung. Following its original casting in a mold built right into the Kremlin's floor, the bell was raised onto an iron grill and remained there for three years while it was cleaned and polished. Prior to its ultimate placement, however, the Tsar Kolokol was ruined during a 1737 fire when cold water was poured on the hot bell, causing a huge section to crack and break off. The bell and the broken piece fell into the original mold in the ground and remained there until 1836. Replicas of the Tsar Kolokol are cast in complete form, with a line to show where the piece broke off; unlike the original, therefore, the replicas are quite capable of producing sound.

Tsar Kolokol bell. This detailed replica of the Great Bell of Moscow was probably cast in bronze using the Lost Wax method. 6.125" high. $250-300.

Here is a another replica of the Tsar Kolokol bell, shown with a ceramic souvenir illustrating the size of the original bell in Moscow.

St. Peter's Basilica in Rome is the home of another widely admired and frequently replicated bell. The original weighs nine tons, and while replicas of the St. Peter's bell vary in terms of their size and the design on their finials, all show the Twelve Apostles encircling the bell's base. Finials may be fashioned to represent the Pope's triple crown, may have one of several kinds of crosses, or may feature an elongated type handle.

Crown and Rose cast pewter replica of the bell from the *Mary Rose,* flagship of Henry VIII's navy, which sank just outside Portsmouth on July 19, 1545. The disc above the handle is in the shape of a tudor rose insignia and shows pictures of the *Mary Rose*. The bottom of clapper is stamped with the Crown and Rose trademark and "MADE IN ENGLAND." 5.5" high. $65-75.

Saint Peter's bell, a replica of the original cast in 1786. This bell is meant to sit on a matching saucer with the likeness of a fallen dove (not shown). The finial is in the shape of the Pope's triple crown. 5.25" high. $300-350 with saucer.

Glazed porcelain replica of the Freedom Bell, which hangs in Berlin's Town Hall. The original bell, eight feet high and weighing ten tons, was given to Berlin by the American people after the airlift that saved the city. The Freedom Bell has been rung every day at noon since its dedication in 1950. The inscription around the bottom reads: "That this world, under God, shall have a new birth of freedom." 3.5" high. $50-75.

Lustrous porcelain replica of the 1936 Olympic Games bell, used at the opening ceremonies for the Olympic Games held in Berlin. Although it never rang again, the bell remains standing as a memorial within the Olympic stadium. Replicas were limited to one per participant during the games. 4.75" high. $100-150.

El Camino Real bells shown with replica mission bell. These tall "guide post" bells are miniature replicas of those found along the seven hundred mile long "highway" that links the California missions founded by Father Junipero Serro between 1769 and 1823. The stand of the left hand bell is inscribed: "MISSION BELL GUIDE POST MARKING EL CAMINO REAL THE KING'S HIGHWAY." The right hand bell is inscribed: "1769-1969 EL CAMINO REAL." The round medallion on the pedestal has a bear in the center and reads: "California Bicentennial 1769-1969." Overall height: 9". Bells: 1.5" high. $20-30 each.

Replica of the new Pummerin. Its predecessor, the old Pummerin, was cast in 1683 after the Turkish Wars and hung in the south tower of St. Stephan's Cathedral in Vienna until it was destroyed in a 1945 fire. The new Pummerin was donated by the province of Upper Austria for St. Stephan's and cast in 1951 using the remains of the old bell. This replica sits on a wooden base with a hole in the center for the bell's clapper. 4" high. $60-75.

Detail of the new Pummerin bell.

Certainly the most celebrated bell in America, and perhaps worldwide, the Liberty Bell is an enduring favorite among bell lovers. First hung in 1753, the bell heralded the 1776 signing of the Declaration of Independence and now stands, albeit silently due to its famous crack, in Philadelphia's Liberty Bell Pavilion. Replicas appear in a multitude of sizes, colors, and materials, but all pay tribute to this famous symbol of American independence. Philadelphia native Charles Boland describes the incredible extent of Liberty Bell copies in his book, *Ring in the Jubilee*:

> Because our Liberty Bell is in its own distinctive way a superlative bell, it is only natural that descriptives applied to it quite logically fall into the superlative. It is the most famous bell in the world (and that simple statement describes it best) and it is the most legend-ridden (and therefore the most misrepresented). It is the most widely-traveled bell in the world and the bell most seen and touched and venerated by the people of the country it symbolizes. There are other "mosts" that can be used in describing it, but perhaps the most astonishing is found in the knowledge that it is the most duplicated, imitated and scale-reproduced bell on the planet. (Boland 1973, 114)

The Liberty Bell has been widely portrayed in two-dimensional as well as three-dimensional form, gracing the front of innumerable medals, plaques, stamps and similar items. Presented here is but a small sampling of the many replicas available to collectors.

Gray colored ceramic Liberty Bell. 3" high. $5-8.

Although replicating an American bell, these two Liberty Bells hail from far away. Left: Wedgwood basalt Liberty Bell, made in England. 3.5" high. $75-100. Right: Sarna Liberty Bell, made in India. 3.25" high. $20-30.

Small carnival glass Bicentennial Liberty Bell with "1776" on one side, "1976" on the other. 2.75" high. $15-25.

Liberty Bells in assorted sizes. Heights range from 3.25" to .75".

Below: Liberty Bells in assorted colors. Heights range from 5" to 3.25".

Figurals and Figurines

Human and animal figures have long been favorite subjects for bells. Bells from long ago and bells of today are universally prized for their depiction of figures, a depiction which can be dramatic, artistic, or even comical in approach. Some collectors specialize in the acquisition of figural and figurine bells alone.

A definition of terms is in order first. Figural bells are those in which the figure serves as the handle of the bell. The figure may stand or sit atop the bell, or may be in the shape of a bust only. A figurine bell is one in which the figure comprises the entire bell, the bell body serving as the skirt or robe worn by the figure. Both types of bells are shown together in this section; each is identified in the caption as either a figural or figurine bell.

The human subjects used for figural and figurine bells typically include historical figures, characters from literature, and "genre" type individuals shown in representative dress or occupation. Given that a skirt or gown most suitably accommodates the shape of a bell—appropriately housing the clapper underneath—figurine bells are almost always in the form of women. The exceptions are men that can be shown wearing a robe, floor-length cloak, or a similar shaped garment. Napoleon, for example, is a male whose apparel lends itself well to the figurine shape and he is seen fairly often in a variety of poses. Similarly, the two male sorcerers shown on pages 117-118 each wear a long, flowing robe that allows their bodies to form the entire bell.

Figural and figurine bells range from those that are highly realistic, with wonderful detailing and faithful reproduction of clothing, hairstyles, accessories, and expression, to those that carry a more primitive or folk art style appearance. The former are most often made of metal, usually brass or bronze, while the latter tend to be found in less durable materials, such as wood or clay. Glass figurines are uncommon. While some may favor one style over another, each has its own unique charm and attraction.

Figurine bell representing costumed ladies of Europe. Many bells of this type were made of women in similar costumes. 4.5" high. $100-150.

Heavy old bronze figurine bell depicting a Belgium lacemaker at work. 5" high. $350-400.

Napoleon is depicted in this figurine bell, shown front and back. The clapper is shaped like a lower leg and foot with a shoe. Men are rarely used for figurines, since their typical attire does not accommodate a bell shape. Two figural versions of Napoleon can also be seen on page 129. 5" high. $50-75.

Here are four different versions of the same figurine bell. This bell is believed to depict either Queen Elizabeth, Mary Queen of Scots, or Queen Victoria; there is not agreement as to which. From left: Bell advertised as Queen Elizabeth with clapper made of two individual feet; bell with clapper of two joined feet; "Queen Elizabeth" bell with plain clapper; simplified Chinese version of the same bell. $15-30 each, depending on workmanship and clapper detail.

Among metal figurals and figurines, it is often difficult to determine the accurate age of the bell or to identify the specific metal used in its manufacture. Old brass tends to soften and darken with age, making it difficult to distinguish from bronze. Most commonly, however, older bells will be significantly heavier than newer ones and the fine quality of the workmanship will be clearly evident. Pay particular attention to the figure's hair and hands, the detailing of which may provide good clues regarding the age and quality of the bell. In addition, the bell's interior may render important, though not always definitive, information about its history:

Metal figurine bell depicting Sally Bassett of Salem, Massachusetts, who was burned at the stake as a witch in 1692. She is shown leaning on her cane. $20-30.

> Patina, dents, and other imperfections of age can all be cleverly faked; but thus far fakers have not created an artificial groove around the inside of the skirt, where the clapper has worn the metal while striking it repeatedly in the same place over the years. The presence of this sign of wear is almost proof positive that the bell is of some age. Its absence, however, does not necessarily mean the piece is of recent manufacture. A genuinely old bell that went unused for a long period may very naturally exhibit an interior as flawless as the day it was made. (Springer 1972, 168)

One of the challenges related to figural and figurine bells—and certainly part of their appeal—is seeking the identity of the individual depicted. Sometimes the intent of the bell's maker is simply to show an "ordinary" person in typical activity; witness the spirited barmaid and the graceful ballerina shown herein. The identity of some of the more common literary or historical bells has been affirmed over the years based on their clear resemblance to other illustrations or descriptions of the same person. Occasionally, however, a bell comes along bearing a figure with such an unusual, unique, or unfamiliar appearance that it almost cries out for investigation of its origin. The answer may appear serendipitously, may come after long searching, or may elude the bell's owner entirely.

Writing in *The Bell Tower*, published bimonthly by the American Bell Association, one bell collector from Fredericksburg, Virginia, described the unexpected, and somewhat ironic, circumstances surrounding her identification of a figural bell that had long intrigued her. The brass bell from her own collection bore a handle in the form of a girl holding a cup in her left hand and a cask under her right arm. She wore a cap, tight-waisted jacket, and trousers beneath her skirt. One day, the local newspaper carried an article about a new Smithsonian exhibit called "Women in Wartime." Smiling out from one of the pictures was a young girl looking just like the mysterious bell. A few museum visits later, the girl from the picture was identified as Mary Tepe, a Civil War "vivandiere" who was awarded the Kearney Cross for bravery after being wounded in the 1861 Battle of Fredericksburg! Vivandieres, the collector learned, were women who assisted with getting supplies such as whiskey and cigarettes to the troops. Their wool uniforms consisted of skirted jackets and trousers, and Mary was especially known for carrying a small keg of whiskey to comfort the wounded soldiers at the front lines. While perhaps not actually depicting Mary Tepe, the figure on the bell almost surely represents one of the Civil War vivandieres and the bell's mystery was happily solved. (Glassco 1995, S18)

Becky Sharp figurine bell. Heavy cast brass or bronze figurine of a mid-nineteenth century lady in a fancy ball gown holding a handkerchief in her right hand and a fan in her left. This bell is universally identified as Becky Sharp, the heroine of Thackeray's *Vanity Fair*. She is a vain, self-centered woman who plays with men's affections and climbs the rungs of society. 4.5" high. $75-100.

Figurine bell representing Sairey Gamp, a character from Charles Dickens' novel *Martin Chuzzlewit*. $35-50.

Reverse of the Sairey Gamp bell.

Typical brass English souvenir bell with flat, one-sided figural handle. This one depicts Mr. Micawber, a Dickens character from *David Copperfield*; many other souvenir bells have been made of Dickens characters as well. "MADE IN ENGLAND" is molded into the interior. 4.625" high. $20-25.

This brass figurine bell is thought to represent the fictional character of the Hag from *The Eve of St. Agnes* by John Keats. 4.75" high. $35-50.

Metal bell with figural handle and raised relief decoration around the base. The woman may represent the "Goddess of Love." 5.5" high. $125-150.

Figural bell with finely detailed bust of Minerva, the Roman goddess of wisdom, mounted on an openwork base. 7" high. $100-125.

Figural bell of graceful but unidentified woman. 7" high. $75-100.

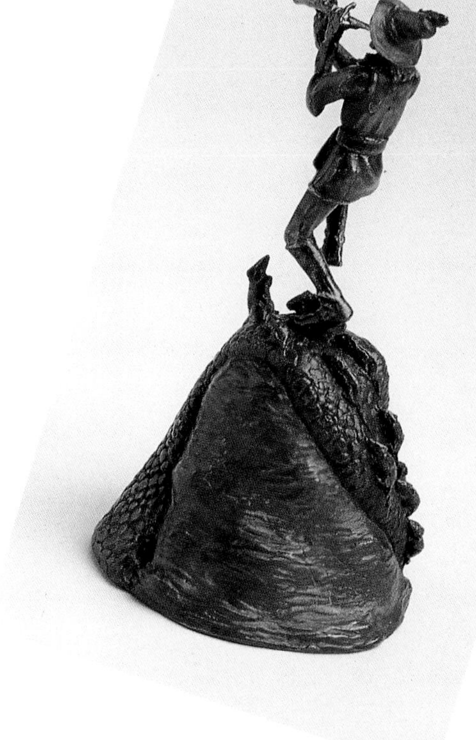

Pied Piper of Hamelin figural bell cast in the Lost Wax method by Blake Valleau of Saugatuck, Michigan. The bronze bell is faintly painted with green on the bell base and the tunic has touches of red. Clapper is a ball on a rod with a rat on the ball. Interior engraving: "Valleau, #015/999." 5.125" high. $235.

Old Majolica figurine made in Italy, known as "The Watermelon Girl." 7.75" high. $200-300.

Unglazed, painted ceramic figurine of a young girl dressed as a red rose, her skirt the opening petals and a red rose clasped in her hand. This was one of three different flower girls available in the gift shop where it was purchased. 4" high. $15-20.

"Church Belle" ceramic figurine from Josef Originals. 3.5" high. $8-10.

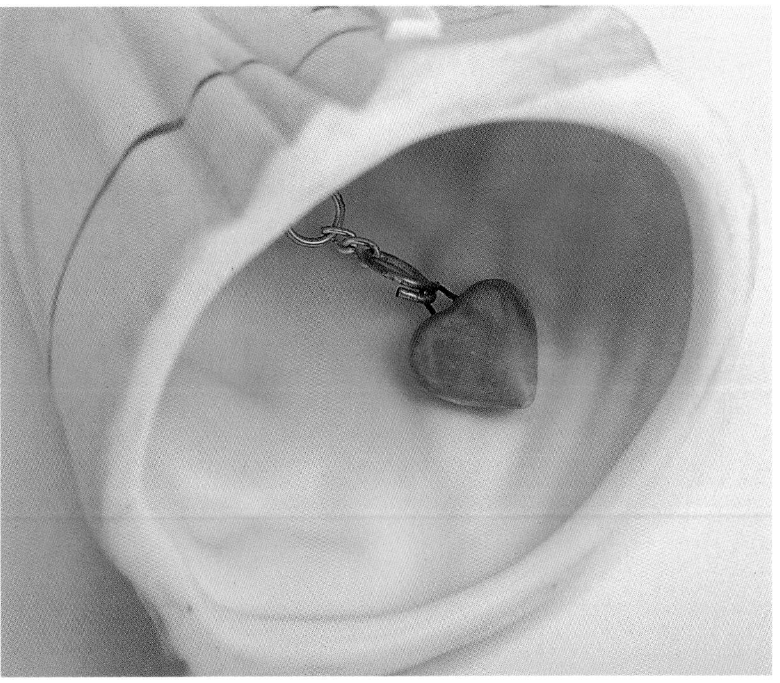

Clapper from the Queen of Hearts bell.

Queen of Hearts bell. Unglazed porcelain bisque figurine, hand painted. The clapper is a pink porcelain heart on a gold chain suspended from a porcelain plug loop. Interior markings identify the bell as made in Mexico and imported by Enesco. 5" high. $15-25.

Flocking provides an interesting texture for these two figurine bells. The Vermont Winter Belle on the left is of hand cast and hand painted ceramic and wears a flocked coat. Her outfit is fashioned after nineteenth century New England dress. 7.5" high. $35-50. "Little Missy" on the right wears a green flocked costume with hip puffs tied at waist. She is also of hand cast, hand painted ceramic. 6.25" high. $20-25.

Heavy glazed Majolica figurine bell depicting a barmaid carrying two mugs of foaming ale in each hand. Markings inside: "1930, 54." 9" high. $250-300.

Close-up of the flocked coat worn by the Vermont Winter Belle above.

Unglazed porcelain Holly Hobbie figurine bell, "A Mother's Memento." Limited Edition, 1979. 3" high. $15-20.

Pair of delicately painted china figurines in pastel colors. Both 4.5" high. $10-15 each.

Mr. Wizard faïence bell. Heavy glazed porcelain figurine in blue design on white; this same bell has been seen in other colors. The clapper is a large blue glazed ceramic bead. 8.25" high. $75-100.

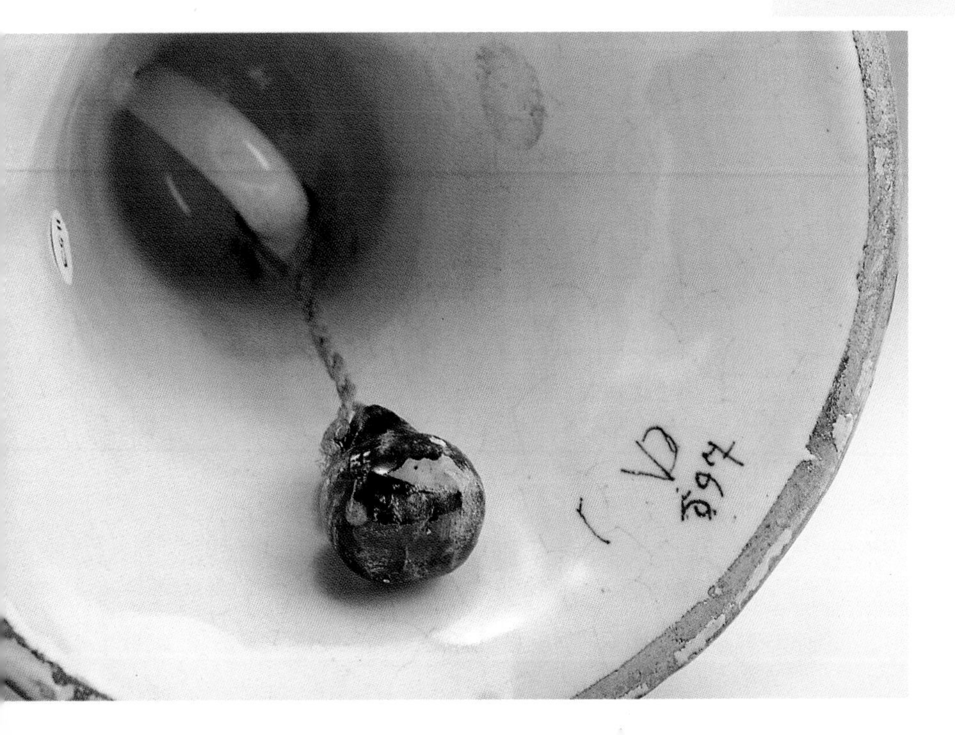

The blue glazed clapper from the Mr. Wizard bell.

Glazed ironstone bell with unglazed bottom rim, depicting "Merlin." Clapper is an unglazed pottery bead on a leather tie. 8.25" high. $30-50.

Bride figurine bell of bone china with skirt made of lattice-type spaghetti work. Made in Taiwan by Enesco. 4.75" high. $35-50.

Another bride, this one of unglazed bisque depicting a colonial woman named "Elizabeth." Signed by the artist on the back, "R. Sauber." 5" high. $40-50.

Stately bride bell by Gorham. Unglazed bisque figurine with "(c) Gorham 1986" printed in small letters inside the skirt. 6.5" high. $40-50.

Folk art style figurine bell of glazed ceramic. 4.25" high. $10-15.

Colorful pair of ceramic figurines depicting Toby and his wife, Agnes. The Agnes bell is more difficult to find. 3" high. $20-25 for pair.

Very simple, almost primitive looking figurine with a wide skirt. The string around the figure's waist is part of the clapper; it threads down through a hole in the back and holds the small pellet clapper inside. Made of lava from Mexico. 4.25" high. $3-5.

Close-up of the figurine's back.

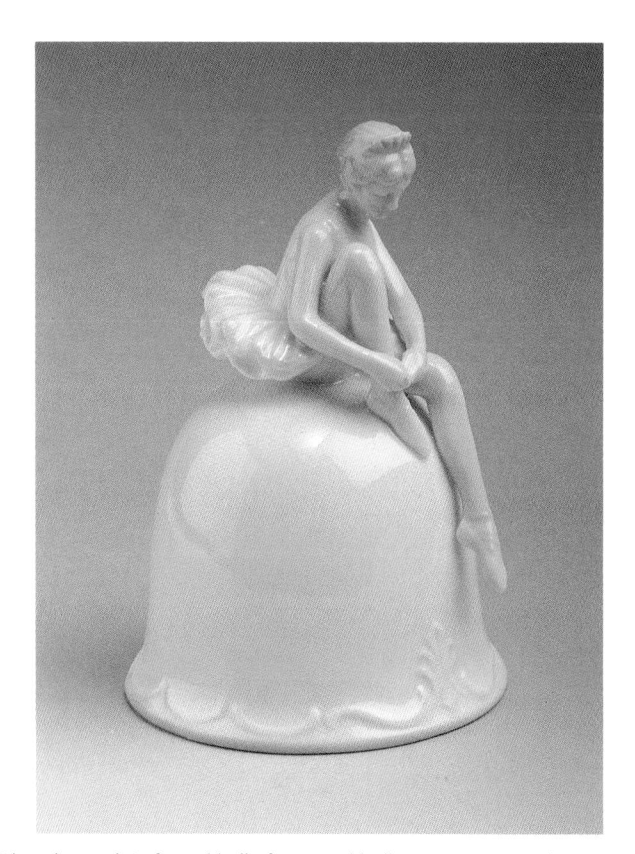

Glazed porcelain figural bell of a seated ballerina in repose. There is only a faint suggestion of color on the figure. The accompanying folder identifies this bell as part of the "Valencia Collection." 4.25" high. $20-25.

Figural bell of small Irish boy seated atop a ceramic bell. 4.75" high. $5-8.

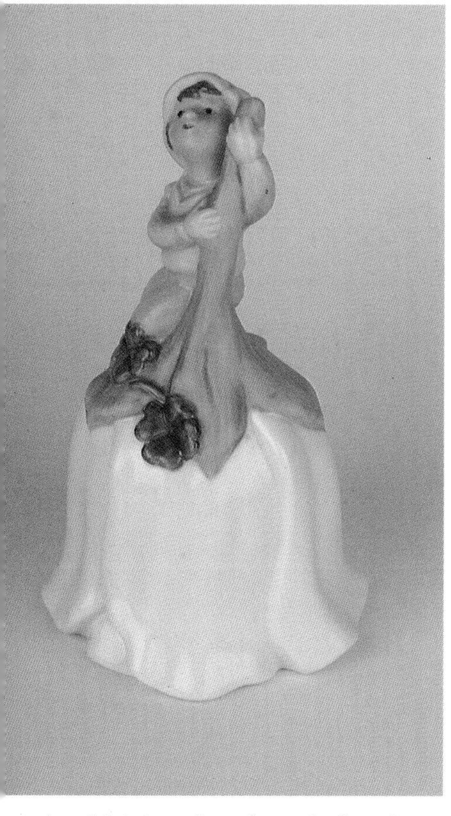

A sprightly leprechaun forms the figural handle on this Good Luck bell from Avon. 4.75" high. $5-8.

Amberina glass figurine of a lady with a flounced skirt and bonnet, made by the Boyd Art Glass Company of Cambridge, Ohio. The skirt is translucent from the waist down, opaque from the waist up. Molded into the lower back edge is a "B" encased in a diamond and the number 2. 4.125" high. $20-25.

White milk glass figurine with crystal glass clapper. 4.5" high. $5-15.

Two glass "Suzanne" bells made by the Imperial Glass Company of Bellaire, Ohio. Both are figurines of a woman in a five-tiered full skirt and low flounced bodice. The bell on the left is molded of iridescent blue glass, the one on the right of iridescent clear glass. The Imperial Glass Company is no longer in business. Both 6" high. $20-30 each.

The Menagerie — Figural Animals

From insects to birds to lions, members of the animal kingdom regularly adorn the top of bells. Majestic to humorous, figural animals are as diverse and assorted as their real life counterparts. Our four-legged and winged friends are certainly not limited to the figural style bell; throughout this book you will find animals portrayed on other kinds of bells as well. Crafted into a handle, however, animals can perhaps most readily be shown in the multitude of poses and positions which so eminently characterize them. So dig out your binoculars and take a stroll through our virtual menagerie!

The handle of this bell is formed by a small bee with outstretched wings. The bell itself is a replica of a nineteenth century cast bronze Russian bell on display at the Metropolitan Museum of Art in New York City. The geometric motif around the base is typical of nineteenth century Russian design. 3.25" high. $25-40.

The National Wildlife Federation issued these two endearing bells. On the left is the Field Mouse bell, featuring a resin molded, hand painted brown field mouse sitting on his haunches on top of the plastic bell base. 4.5" high. Next to him is the Rabbit Family bell, a hand painted group of three rabbits atop a clear lead crystal bell. 4" high. $25-35 each.

Pair of two brightly painted wooden bells with animal handles and wooden clappers. Frog: 6" high. Bird: 6.5" high. $5-8 each.

Blue jay bell of fine bone china by Towle. The bell base is designed to resemble tree bark. 5" high. $15-20.

Handmade pottery parrot bell from New Jersey. 6" high. $30-40.

Detail of the blue jay's expressive face.

Lenox chickadee bell. Glazed porcelain, with finely painted, unglazed black-capped chickadee on top. Inside markings: "FINE PORCELAIN, LENOX, 1991, Black-capped Chickadee." 4.125" high. $50-65.

Two metal bells of unknown origin, both depicting birds on their handles. Left: 3.25" high. Right: 5" high. $5-15 each.

Great horned owl bell, part of a National Audubon Society collection.. Glazed china bell with porcelain bisque owl on top. Inside is the logo of the Audubon Society and "Great Horned Owl, (c) 1990 National Audubon Society, Licensee Enesco Corporation, Made in Malaysia". Overall height: 6". $50-60.

Zebra head bell, also of glazed china. 4.75" high. $30-50.

A jaguar's head serves as the handle for this spotted bell of glazed china. 5.5" high. $30-50.

The lion on this ceramic bell wears a soulful expression. His mane and the base of his tail are made of strands of "spaghetti-like" clay hairs, sticking out every which way. 4.75" high. $25-30.

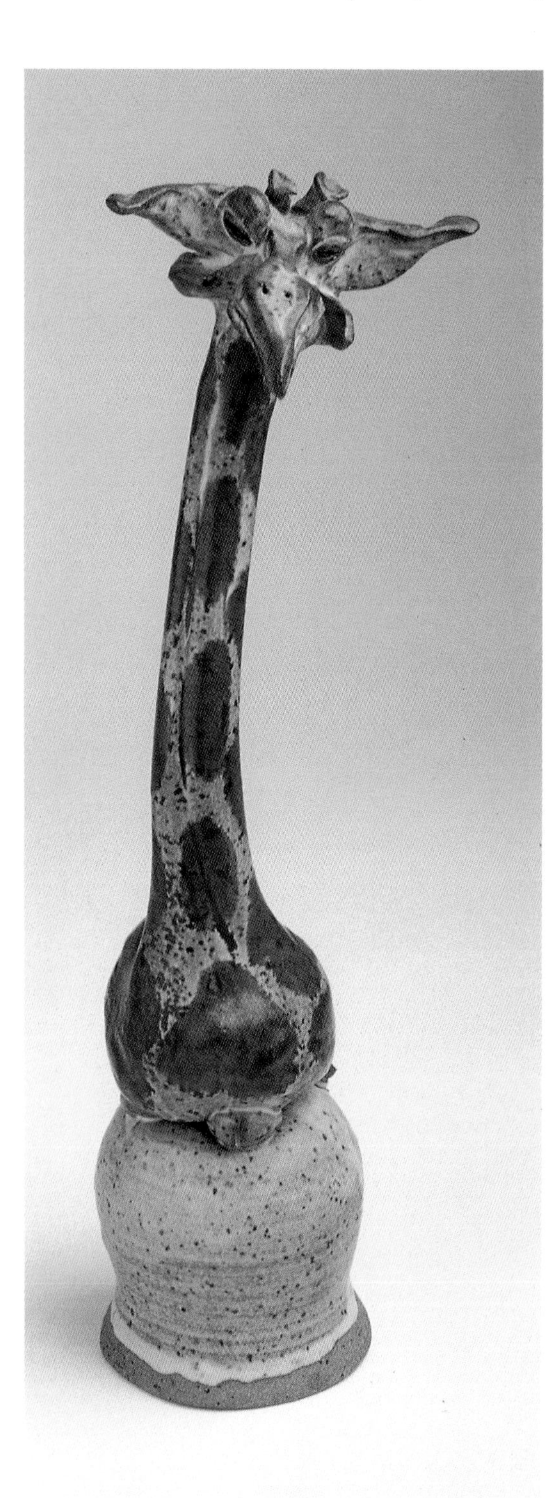

A mellow-looking giraffe sits atop this ironstone bell, forming its "handle." The owner of this bell would certainly want to place it in a secure location where it would not be easily knocked over! 15" high. $50-75.

Detail of the giraffe's head.

Pair of white ceramic bells with matching figural handles in the shape of happy monkeys. 4.25" high. $8-10 each.

Although not truly figurals, these two comic bells deserve a place in our menagerie. The hand-crafted cat on the left looks appropriately smug as he contemplates the look of anxiety on the mouse sitting next to him. Purchased at the Grand Bazaar in Istanbul, the mouse has a clapper consisting of a round clay bead which shows through his wide open mouth. $10-15 each.

Well, it's some kind of animal...! This whimsical bell is partially glazed with a cream colored glaze and horizontal striations. Atop stands a creature looking like a rather sad cow masquerading as a dragon. 8" high. $20-30.

Historical People and Events

Well-known figures and events from the annals of history have long been favorite subjects for bells. Along with their appearance in paintings, sculpture, and photography, many famous people have been immortalized on the handle or body of a bell. Two shown here, Napoleon and Joan of Arc, are among those most prevalent. In addition to actual people, figures representative of a particular time or event may be found on bells; the Roman centurion and Revolutionary War soldier are examples. While figural or figurine bells are naturally the most typical kind used for illustrating famous people or events, other styles are seen as well.

Cast bronze bell depicting a Roman Centurion, manufactured by Emil Huddy of California. The hexagonal bell base has a simulated Roman coin on each face. On top stands a Roman centurion in traditional garb with a long spear in his right hand. Scratched on the bell's interior is "Huddy, 1985, Ser. #161." 7" high. $75-100.

Sennacherib's Army or Assyrian bell, brass. The relief around the body of the bell shows Assyrian warriors carrying heads of Hittite captives after a battle. From a seventh century relief in London's British Museum. 5" high including handle. $100-125.

Detail of the relief around the Sennacherib's Army bell.

Another figural bell of Napoleon, this one with bas relief scenes from the famous Battle of Waterloo in 1815. 5" high. $35-50.

Heavy bronze figural bell with Napoleon on handle. Napoleon holds one arm behind his back, a spyglass in the other. On a band above the base "SOUVENIR" is incised above an eagle and "WATERLOO" above a lion. 5.25" high. $35-50.

Joan of Arc figurine bell. Heavy cast bronze of a full-skirted, armor-clad Joan with arms crossed against her breast. Since her head is uncovered, it is generally felt that the object in her right hand may be her helmet. 4" high. $35-50.

Left: Heavy old English bell of cast brass with a two-sided bust of Charles Edward Stuart, also known as "Bonnie Prince Charlie" (1720-88). Prince Charles led the Scottish Army in an uprising of the clans to restore Stuart kings to the throne of England. Culloden Moor is the battlefield in Scotland where his troops were badly defeated, essentially ending this 1745-46 rebellion. 4" high. $40-50.

Figurine bell of Marie Antoinette, made of heavy bronze. She holds her left hand at her breast and a butterfly net in her right hand. 5.5" high. $125-150.

Two bells from a series of six called "Women Who Changed History," produced by the Hamilton Collection and made by the Gorham Manufacturing Company. On the left is Queen Isabella, of cast bronze coated with Gorham silver, wearing a long parted gown and carrying an open Bible. On the right is Queen Catherine of Russia, silver-plated with finely detailed casting. Catherine wears a full skirted gown embossed with coat of arms. The cloak over her shoulders falls to the base in back with embossing that suggests ermine tips around the edge. Both 5" high. $75-100 each.

Gold plated figural bells depicting four queens of England. From left: Mary I, 1553-1558; Elizabeth I, 1558-1603; Mary II, 1689-1694; Victoria I, 1837-1901. Each bell has the name, date, and designer's name (Procopio) incised in the mold of the bust. 3.75"-4" high. $25-40 each.

Pair of English bells honoring Benjamin Beale, an Englishman who in 1753 invented the "bathing machine on wheels," and Martha Gunn, a wash woman who assisted with the women and children using the machine. 3.5" high. $25-30 each.

Figurine bell of Florence Nightingale, famous British nurse who oversaw nursing efforts during the Crimean War and who spearheaded the start of professional education for nurses. 5" high. $75-100.

Figurine bell of Clara Barton, American humanitarian who volunteered during both the American Civil War and the Franco-Prussian war and was the founder of the American Red Cross in 1881. 5" high. $75-100.

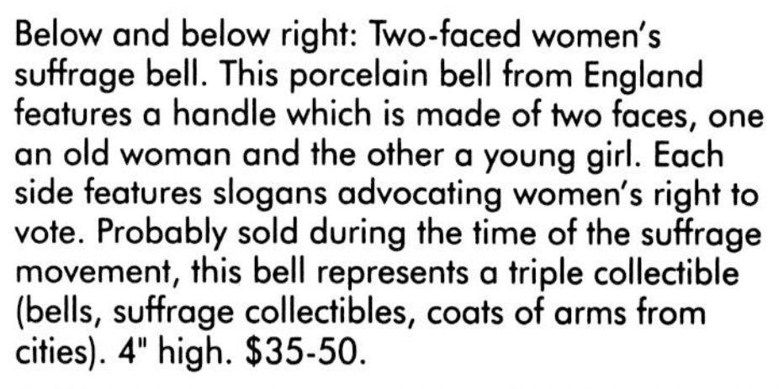

Below and below right: Two-faced women's suffrage bell. This porcelain bell from England features a handle which is made of two faces, one an old woman and the other a young girl. Each side features slogans advocating women's right to vote. Probably sold during the time of the suffrage movement, this bell represents a triple collectible (bells, suffrage collectibles, coats of arms from cities). 4" high. $35-50.

The plain, silver-plated base of this bell is surmounted by a finely detailed pewter figure of a Revolutionary War soldier, representing one of the beleaguered men who braved the winter of 1777 at Valley Forge. Inscribed inside: "THE LINCOLN MINT, 1976, 1552." 4.75" high. $20-25.

Two bells honoring the Bard of Avon, William Shakespeare. The figure of Shakespeare on the left stands atop a standard souvenir base used for many similar English bells. 5.75" high. $60-75. The flat, one-sided handle of the bell on the right portrays Shakespeare's birthplace. 5.5" high. $8-10.

These two bells are from a series depicting American first ladies. On the left is Dolly Madison, a hand painted bisque-fired figurine. A paper label inside reads: "Gorham Made in Japan, #683." 5.5" high. On the right is Martha Washington, also of unglazed bisque-fired porcelain. She has the same label inside, except that her number is #1270. 5.375" high. $30-40 each.

Left: Figural bell with bust of William II (also known as Kaiser Wilhelm), emperor of Germany and king of Prussia from 1888 to 1918. 4.5" high. $20-25.

Fenton "Famous Women" bell, limited edition from the Connoisseur Collection. Ruby satin iridescent glass bell with three portrait cameos of accomplished American women. Right: St. Elizabeth Ann Seton. Below: Helen Keller. Below right: Amelia Earhart. Clapper is a typical Fenton clear glass bead on a gold chain. 6.25" high. $20-30.

Series, Annuals, and Limited Editions

As with other types of collectibles, numerous companies and individual artists have created bells which are issued on an annual basis, as a series, or as limited editions. Acquiring a complete set of these bells can be a highly desirable goal, and even a partial set makes an attractive display to be admired and appreciated on a daily basis.

Flowers, birds, figurines, and Christmas bells are popular subjects, but bells in a series are far from limited to these areas. Some of the manufacturers that have issued limited edition or series bells include the Danbury Mint, Enesco, Fenton Art Glass, Goebel, Inc., Gorham, Lenox, Schmid, and Towle Silversmiths.

Many of the series and limited edition bells available date from contemporary times, although this does not necessarily decrease their value or make them less collectible. One well-known and highly sought set was issued on two different occasions; these are Royal Bayreuth's Sunbonnet Baby bells. The design for these bells is credited to artist Bertha L. Corbett, who originally created the faceless but charming figures for a book published in 1900. The days of the week series shows the Sunbonnet Babies engaged each day in a different household chore. Early in the twentieth century, Royal Bayreuth purchased the rights to Corbett's designs and used them to illustrate a series of porcelain bells. The series was reissued in 1977 in a limited edition of 1500, slightly smaller than the original set but otherwise identical. Two of the 1977 bells are shown on page 141.

Three bells by Peter Barrett, part of a series of bells with bird themes produced by the Franklin Mint. The birds are realistically detailed and virtually life-size. From left: Goldfinch, 4.75" high; nuthatch, 5.375" high; wren, 5.125" high. The clappers on all three bells are unglazed white oval eggs. $50-75 each.

Four porcelain bells representing characters from fairy tales. From left: Snow White, 6" high; Cinderella, 5.75" high; Rapunzel, 6" high; Sleeping Beauty, 4.375" high. All are Franklin Porcelain bells sculpted by Trina Hyman. $75-100 each.

The bells in this group of colorfully painted ceramic caricatures are known as "Daffy Bells" and date from the late 1950s. Marked Japan; average height, 5". $5-10 each.

Cigar smoking "Daffy Bell" man in derby hat.

Blonde "Daffy Bell" with top knot and pearls. Dressed for a night out!

Pewter series of figural bells showing Jesus Christ with the twelve Apostles. This set was made by Englefield's of London and each bell has the Crown and Rose mark on its pewter clapper. All 4.75" high. $40-50 each.

Jesus and the Apostles Thaddeus and Simon.

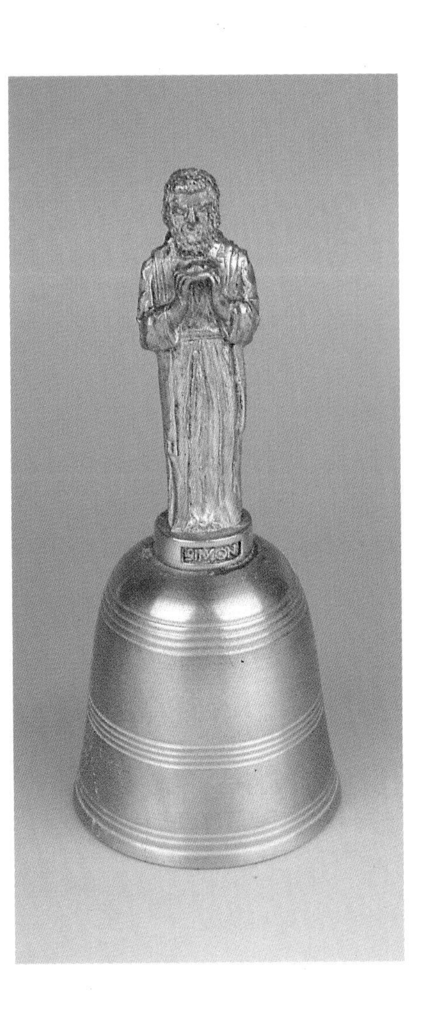

World of Children series by River Shore Ltd. Porcelain glazed bell bases with unglazed figures on top. Food representative of each child's country rims the bowls. From left: "Katrina," 1980, Netherlands; "Allison," 1979, United States; "Rei-Ling," 1979, China; "Kulak," 1980, Arctica. All 7.25" high. $30-35 each.

Close-up of "Allison" bell from the World of Children series.

Christmas ornaments issued by the Salamander Pottery of Port Orchard, Washington. These have been issued each year since 1980 and each bell is signed and dated. Back row, from left: Clown, 3" high, 1987; Miss Hedgehog, 3" high, 1996; Toucan, 3" high, 1994. Front row, from left: Dragon, 2.625" high, 1984; Green Frog, 1.5" high, 1995. $15-25 each.

Above: Bells of the World series. Porcelain glazed bells with overpainting of faces and other details. Back row, from left: Japan, Greece, Scotland, China. Front row, from left: Sweden, USA, Israel, Germany, Ireland. Each bell has a paper label inside with the country's name and "SCHMID BROS. INC. Made in Japan." Heights range from 5.25" to 5.5". $10-15 each.

Below: Two bells from a series representing months of the year and their special flowers. Left: "February" is a hand painted figurine made of unglazed porcelain bisque. Inside is marked: "February, Christabelle with Violets, (c) 1986 Enesco Imports Corp., Made in Mexico." Right: The "March" figurine holds a spray of white and yellow daffodils. Inside markings are the same except for the name: "March, Victoria with Daffodils." Both 5" high. $20-25 each.

Two bells from a series of 1993 bells called "A Christmas Carol," by Marjorie Sarnat. Left: Ebenezer Scrooge's House, made of unglazed, hand painted, porcelain bisque. Right: Bob Cratchit's House. Both 5" high. $15-20 each.

One of a series of twelve showing different buildings, this Village Cottage bell from the Lenox Porcelain Company is crafted in the shape of a house. 4.25" high. $20-25.

Lenox trademark inside the Village Cottage bell.

Sunbonnet Baby bell, "Monday." Glazed porcelain bell with gold "trefoil" crown made by the Royal Bayreuth Company of Germany. This bell is from a set of reproductions made c. 1977 and is marked "No. 910 of a Limited Edition of 1500." Monday represents washing. 3.5" high. $30-40.

Sunbonnet Baby bell "Wednesday," representing mending. Also from the set of reproductions and marked "No. 140 of a Limited Edition of 1500." 3.5" high. $30-40.

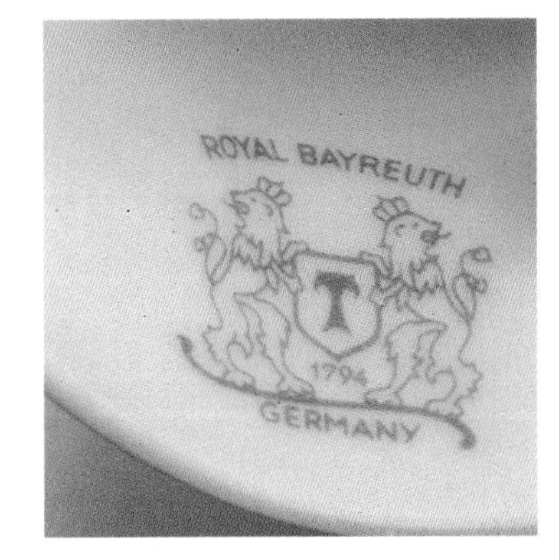

Royal Bayreuth coat of arms trademark inside the "Wednesday" bell.

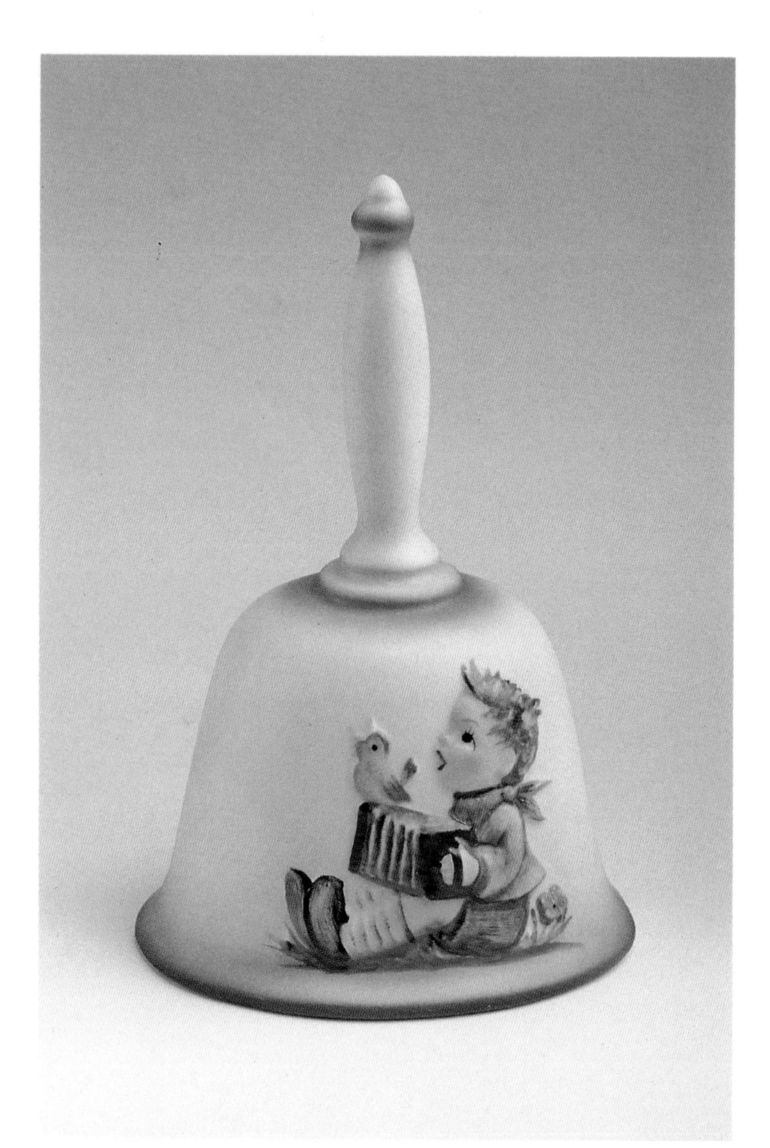

This 1978 annual Hummel bell was the first in a long series by Goebel of West Germany, each with a different Hummel figure. Unglazed, cream colored bell with Hummel figure in bas relief on one side, 1978 in red on the other. This is a First Edition called "Let's Sing." 6.25" high. $35-50.

1978 First Edition porcelain bell from a series called "Bells of the Four Seasons." The red cardinal represents winter. 6" high. $150-200.

The works of two twentieth century American sculptors have become especially prized for their quality, artistry, and creativity. The first is a series of impressively designed animal bells sculpted and cast in Colorado by Carl Wagner. Each limited edition bell depicts the head or body of a different animal, including an elk, elephant, ram, and horse. Wagner cast these heavy, expressive bells using the time consuming Lost Wax method; each has a corresponding clapper which simulates the foot, hoof, or paw of the respective animal. The clappers and an additional Wagner bell are shown in Chapter Six.

Wapiti elk bell, by Carl Wagner. 5" high. $300-400.

Elephant head bell, by Carl Wagner. Very heavy brass bell with realistic clapper in the shape of an elephant's leg and foot (clapper shown in Chapter Six). Incised inside: "C. Wagner (c) 78, 29/500." 6.25" high. $300-400.

Bellmaker Gerry Ballantyne of Overland Park, Kansas, the other American sculptor, is considered one of the most skilled and creative artisans in the world of bells today. His extraordinary series of figural bells, also cast in the Lost Wax method, are among those most fervently sought by current collectors. Active in bells since 1953 and a former president of the American Bell Association, Ballantyne made his first bell in 1972. He originally issued one annual and one special edition bell each year; the bells proved so popular, however, that he found it necessary to cut back on this two-a-year schedule.

Augusta Livia Drusilla. Mother of Tiberius Nero, the second emperor of Rome. 1972 annual bell by Gerry Ballantyne. 6" high. $750-800.

Maria Theresa (1717-80). Archduchess of Austria, queen of Austria and Hungary, and mother of Marie Antoinette. 1973 annual bell by Gerry Ballantyne. 6" high. $750-800.

Hans Brinker. Young Dutch character from *The Silver Skates*, by Mary Dodge. 1974 annual bell by Gerry Ballantyne. 5.75" high. $400-500.

The highly detailed figures on each of the Ballantyne bells come from both history and literature; their fascinating stories account for part of the bells' attraction. Each bell is issued in a limited edition of 200 to 400, with most averaging 250. The individual bell's theme is often carried onto the decorative base: battle scenes below Joan of Arc, dragons below the Chinese Mandarin, a pumpkin coach below Cinderella. Several of the bells are nodders, meaning that their heads or feet move rhythmically when pushed, and all have clappers inscribed with the letter "B" enclosed by a circle.

Paul Revere (1735-1818). American silversmith, engraver, and patriot, whose famous "midnight ride" of April 18, 1775 made him a folk hero of the Revolutionary War. 1976 annual bell by Gerry Ballantyne. 5.75" high. $500-600.

Catherine of Aragon (1485-1536). Daughter of Ferdinand and Isabella, the Spanish monarchs who sponsored Columbus on his voyage to the New World, and first of the six wives of England's Henry VIII. 1975 special edition nodder bell by Gerry Ballantyne. 4.75" high. $750-800.

Henry David Thoreau (1817-62). American writer, philosopher and naturalist, who lived in a hut by famous Walden Pond from 1845 to 1847. 1975 annual bell by Gerry Ballantyne. 6" high. $300-400.

Bicentennial Eagle, commemorative of America's 200th anniversary. 1976 special edition bell by Gerry Ballantyne. 6" high. $200-300.

Becky Thatcher. Fictional character from *The Adventures of Tom Sawyer,* by Mark Twain. 1977 annual bell by Gerry Ballantyne. 5.75" high. $400-500.

Jesus Christ, by Gerry Ballantyne. Open stock through 1989. 5.75" high. $200-300.

Joan of Arc (1412-31). Born to peasants, she led the French to several military victories against the English in 1429. She was burned at the stake for heresy in 1431 but later pronounced innocent and canonized as a saint. 1979 annual bell by Gerry Ballantyne. 6" high. $150-200.

The Gibson Girl. Well known fashion icon originally conceived by American illustrator Charles Dana Gibson in the late nineteenth century. 1978 special edition bell by Gerry Ballantyne. 6" high. $150-200.

Mandarin Indian Buffalo Dancer. Dancer wearing buffalo head as part of ritual dance to attract buffalo herds that were used by Indians for their meat and hides. 1978 annual bell by Gerry Ballantyne. 5.75" high. $200-300.

Friar Tuck. One of the legendary comrades of Robin Hood, helped to steal from the rich and give to the poor. 1979 special edition nodder bell by Gerry Ballantyne. 4.75" high. $350-400.

The Chinese Mandarin. A high-ranking military or civil official belonging to one of nine ranks during the period of the Chinese Empire. The official's rank determined the color of the button worn on his hat. 1980 annual bell by Gerry Ballantyne. 6" high. $100-150.

Reverse of the Chinese Mandarin.

Evangeline. Tragic heroine of Henry Wadsworth Longfellow's poem by the same name. The story describes the plight of two lovers separated during the eighteenth century French and Indian War. 1980 special edition bell by Gerry Ballantyne. 5.75" high. $150-250.

The Pioneer Schoolteacher. Prototypical young schoolteacher of the late nineteenth century, carrying tools of the trade: a precious book and a brass school bell. 1981 annual bell by Gerry Ballantyne. 6" high. $150-250.

Robin Hood. Reputed "gentleman" outlaw of Sherwood Forest, the subject of many old English ballads. While the authenticity of his existence is not truly known, his colorful life has been widely romanticized. 1981 special edition bell by Gerry Ballantyne. 6.5" high. $150-250.

The Court Jester. Traditional joker and fool from old England, expected to amuse and entertain the royal family and nobility. Small bells adorned the jester's hat and clothing, tinkling as he danced. 1982 annual bell by Gerry Ballantyne. 6" high. $150-250.

The Town Crier. Essential purveyor of community information in the days before newspapers, wearing his characteristic breeches and tri-cornered hat and carrying the all-important bell. 1982 special edition bell by Gerry Ballantyne. 6" high. $150-250.

The Circus Clown. Beloved entertainer who uses makeup, costume, and antics to generate laughter from his audience. 1983 annual bell by Gerry Ballantyne. 5.75" high. $150-250.

La Esmeralda. Gypsy heroine from *The Hunchback of Notre Dame*, by French novelist Victor Hugo. Saved temporarily from execution by Quasimodo, the bell ringer of Notre Dame. 1983 special edition nodder bell by Gerry Ballantyne. 5.25" high. $275-350.

Cinderella. Heroine from the well known children's story whose fairy godmother and lost glass slipper changed her life. 1984 annual bell by Gerry Ballantyne. 6" high. $150-250.

Reverse of Cinderella.

The Swiss Herdsman. Alpine dairyman who takes great pleasure and pride in tending his herd of contented bovines. 1984 special edition bell by Gerry Ballantyne. 6.25" high. $150-250.

Della. Devoted and unselfish wife from O. Henry's classic *Gift of the Magi*, who sold her hair to buy husband Jim a platinum watch fob for Christmas. 1985 annual bell by Gerry Ballantyne. 6.25" high. $150-250.

James Bridger (1804-81). American fur trapper, explorer, and scout. Hunted in the Rocky Mountain region and provided early descriptions of natural phenomena in what is now Yellowstone National Park. 1985 special edition bell by Gerry Ballantyne. 6" high. $150-250.

Jack and the Beanstalk. Colorful tale describing Jack's fateful purchase of magic beans and his subsequent encounter with a giant at the top of the beanstalk. 1986 annual bell by Gerry Ballantyne. 5.75" high. $150-250.

Omar Khayyam (c. 1050-1152). Persian mathematician, astronomer, and author of *The Rubaiyat*, one of the world's most famous works of poetry. 1986 special edition nodder bell by Gerry Ballantyne. 4" high. $150-250.

Dorothy. Spunky heroine who searches for *The Wizard of Oz* in Frank Baum's famous 1900 story, shown with faithful dog Toto. 1987 annual bell by Gerry Ballantyne. 6" high. $150-250.

Reverse of Dorothy and Toto.

Rip Van Winkle. Long-bearded fictional character from Washington Irving's 1819 story by the same name, who fell asleep in the woods for twenty years. 1987 special edition bell by Gerry Ballantyne. 6.25" high. $150-250.

Edgar Alan Poe (1809-49). American poet and writer, considered a master of the macabre and suspenseful short story. Among his famous poems is an eloquent ode to *The Bells*. 1988 annual bell by Gerry Ballantyne. 6" high. $150-250.

Mumtaz Mahal. Wife of Mughal Emperor Shah Jahan and inspiration for the architecturally classic Taj Mahal, which serves as her tomb. 1988 special edition bell by Gerry Ballantyne. 6.25" high. $150-250.

Reverse of Mumtaz Mahal.

Aladdin's Genie. Fabled genie released by Aladdin to grant his every wish in the romantic folktale known as *Aladdin and the Enchanted Lamp*. 1989 annual bell by Gerry Ballantyne. 6" high. $150-250.

Reverse of Aladdin's Genie.

Santa Claus. Mythical and revered legend of Christmas, originally derived from an actual Catholic bishop named Saint Nicholas. 1989 special edition nodder bell by Gerry Ballantyne. 5.5" high. $250-350.

Empress Eugénie (1826-1920). Born in Spain and educated in Paris, she married Napoleon III in 1853 and served as empress of France until 1871. 1991 annual bell by Gerry Ballantyne. 6" high. $150-250.

Cleopatra (c. 69-30 BC). Egyptian queen, famous for her beauty and ill-fated love affairs with Julius Caesar and Mark Antony. 1990 special edition bell by Gerry Ballantyne. 6" high. $150-250.

Ponce de León (1460-1521). Spanish explorer and conqueror, accompanied Christopher Columbus on his second voyage to America and searched in vain for the reputed Fountain of Youth. 1990 annual bell by Gerry Ballantyne. 6.25" high. $150-250.

Blackbeard. English pirate of the early eighteenth century, infamous for his robberies throughout the West Indies and American colonies and for his harsh treatment of resistant prisoners. 1992 annual bell by Gerry Ballantyne. 6" high. $150-250.

Hopi Kachina. Living in the barren mesas of North Arizona, Hopi Indian men dress up each year as Kachina dolls, which they believe are the embodiment of the natural world, as well as their partners in helping prepare the world for each growing season. 1993 annual bell by Gerry Ballantyne. 6" high. $150-250.

The Scarlet Letter. Depicts Hester Prynne, tragic character from Nathaniel Hawthorne's 1850 novel, who endured public condemnation for giving birth to an illegitimate child by the town pastor. 1994 annual bell by Gerry Ballantyne. 6" high. $150-250.

Sarna Bells

The son of an Indian dairy farmer, Sajjan Singh (S.S.) Sarna's name has become almost synonymous with bells and his long term marketing strategies for "The Bells of Sarna" have earned him a permanent spot in the annals of bell collecting lore.

Sarna was born in the late nineteenth century and initially educated at a Presbyterian missionary school in India. He left home at nineteen to escape a prearranged marriage and ultimately found his way to the United States. There he attended college, planning to return home and work with his father in the dairy industry. To earn extra money at college, Sarna gave talks on Indian life and culture to other students. Surprised by the intense interest his friends had in Indian handcrafts, Sarna abandoned his college plans and, with his father's assistance, became a budding entrepreneur. He tried many different goods, achieving limited success, until a hunch in 1933 led him to borrow money, travel to India, and purchase an initial supply of various bells. (Schick 1981)

Returning to America, Sarna found that prospective buyers wanted to know how the different bells were used. To satisfy their curiosity—and dramatically increase his sales at the same time—he began attaching small descriptive story tags to each of the bells. Sarna later acknowledged that while many of his bells were used for specific purposes in India, they were not really known by any special name; the names and stories found on the tags, therefore, were somewhat improvised to help with successful promotion of the product. (Schick 1981)

And most successful it was! By the late 1930s more than a million "Bells of Sarna" were being sold each year and overall sales in the early 1960s reached fifty-five million. Sarna's hunch had repaid him handsomely. (Schick 1981)

Sarna bells are amazingly diverse. Most commonly found are those strung together as a set, typically accompanied by the standard Sarna identification tag and an informational booklet. Additional bells found in his catalogs include "special bells with conventional handles, figural handles, wind bells and bells on chains, straps, braided cords, brackets, cymbals and gongs, fabric panels, and limited editions such as the Christmas Bells and Bicentennial Bells...miniatures, baby bells, wedding bells, prayer bells, sweetheart bells, zodiac bells, animal bells..." (Schick 1981)

Sarna was an active member of the American Bell Association (ABA) and promoted the organization through the booklets attached to his bells. A page in the original booklet gave a description of the association and recommended that beginning collectors join the group. The membership fee was listed as $1.75 and a name and address to write to was provided. Reportedly, the contact person named (still a member of ABA) receives occasional membership requests to this day based on Sarna's tag information. The membership fee has naturally increased in accord with the times but the remaining ABA information from the booklet remains quite accurate!

S.S. Sarna died in 1978. A selection of his annual Christmas bells can be found in the section on Seasonal and Holiday bells.

One of the well-known Sarna booklets which accompanied many of his bells.

String of four Sarna bells. This string has a 1955 "Bells of Sarna" booklet but no individual bell tags. Top to bottom: Curry Kothi (Curry Ritual) bell; Holi (Water Festival) bell; Kohlu (Oil Press) bell; Najoomi (Fortune Teller) bell. $10-20.

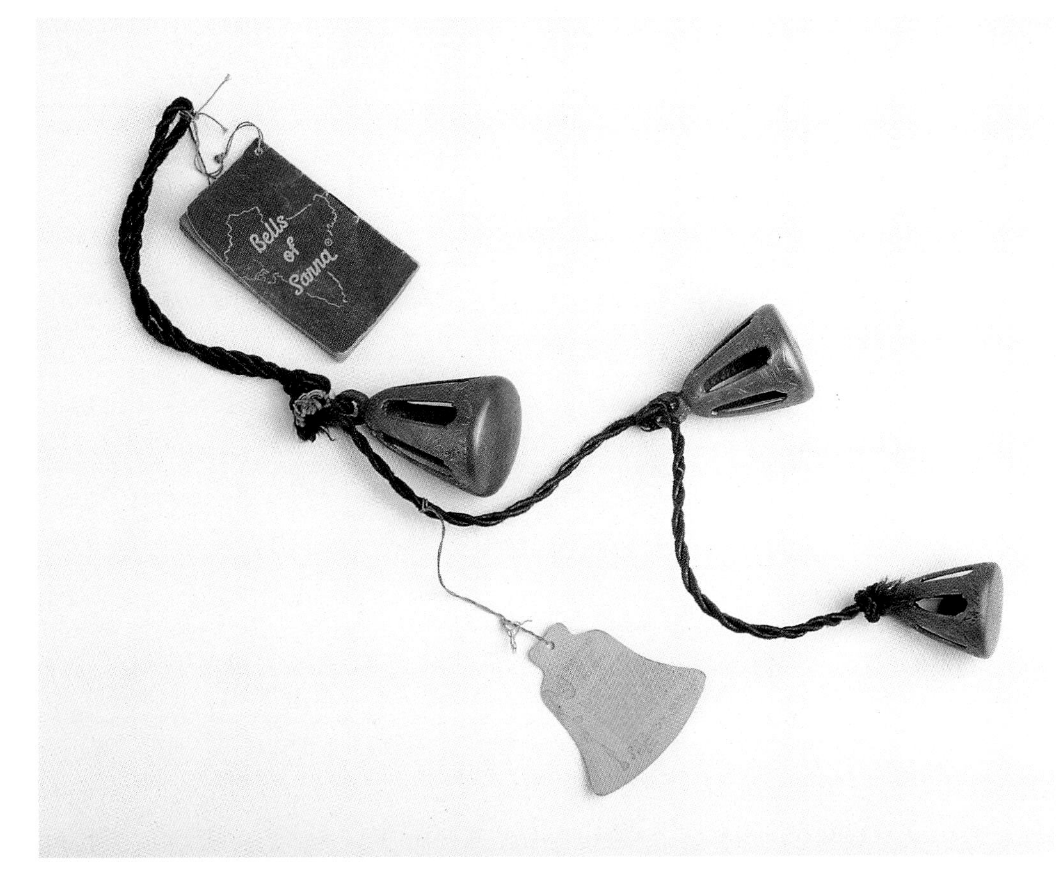

Dak-Runner (Mail Carrier) bells. Old string of three graduated cone-shaped crotal bells on a black cord with original Sarna booklet and tag attached. Booklet has a 1955 copyright date. $10-20.

Three bell Sarna string of Sacred Pipal Tree bells joined by a brown and yellow silk braided strap. Also on the string are the original tags, a "Bells of Sarna" booklet, and a tag printed in India with a 1949 copyright date. $10-20.

Hawker bell, made by Sarna. The tag describes this bell as one typical of those used by peddlers in Java: while walking the streets to sell their wares, the peddlers ring these bells constantly. Each peddler's bell has a distinctive tone. Engraved on side: "SSS, 530, JAVA." 3.5" high. $10-20.

String of Anarkali or Pomegranate bells by Sarna. Four "pomegranate" shaped crotals all fastened to a typical Sarna twisted strap. No Sarna markings. $10-20.

String of three Sarna Sweetmeat bells, with original tag and 1955 booklet. According to the tag, sweetmeat shops in India feature tasty delicacies arranged on plates that encircle the shop's owner. Customers who desire items far from their reach use bells hanging from the ceiling to keep their balance while they lean over to obtain the sweet. $10-20.

Miniatures

One of the most fascinating aspects of bells is their size range, running from virtually mammoth to nearly minute. All are interesting and admirable in their own right, but while the world's largest bells can generally be appreciated only through pictures, visits, or small-scale replicas, those of a more diminutive nature are far more practical to include in a collection! There is no specific guideline for what qualifies a bell as "miniature;" a height of two inches or less might be a good rule of thumb. Those shown here all meet that standard, and bells of an even tinier size can certainly be found.

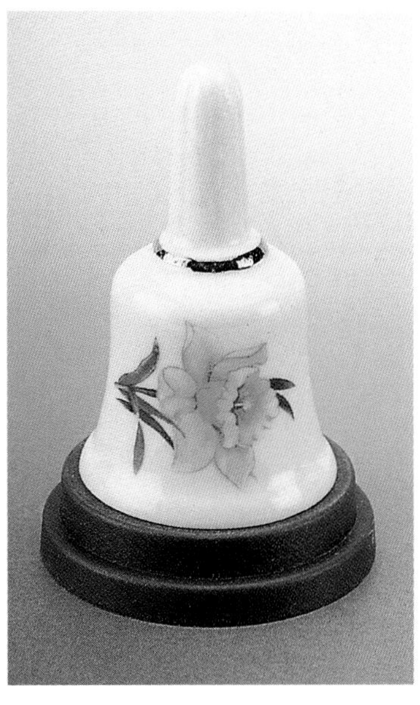

Tiny daffodil bell of fine bone china comes from Wales and has its own green stand. 2" high without stand. $5-8.

Realistically molded and decorated, these four painted pewter bells are less than 2" high each. $15-20 each.

Petite brass cowbells purchased in Bern, Switzerland. 2.75" high including strap. $3-5 each.

Only 1.5" high, this Royal Tara bone china bell is delicately painted with Irish shamrocks. $3-5.

Left: Three miniature bells shown with larger bell of similar design for comparison in height. Large bell: 5.25" high. Small bells: 1.5" average height. $2-5 each for miniatures.

Advertising Bells

Some use signs, some use newspaper, some use...bells! Close cousins to the souvenir bell, these bells used for advertising take advantage of attractive or eye-catching designs to promote a specific product or place.

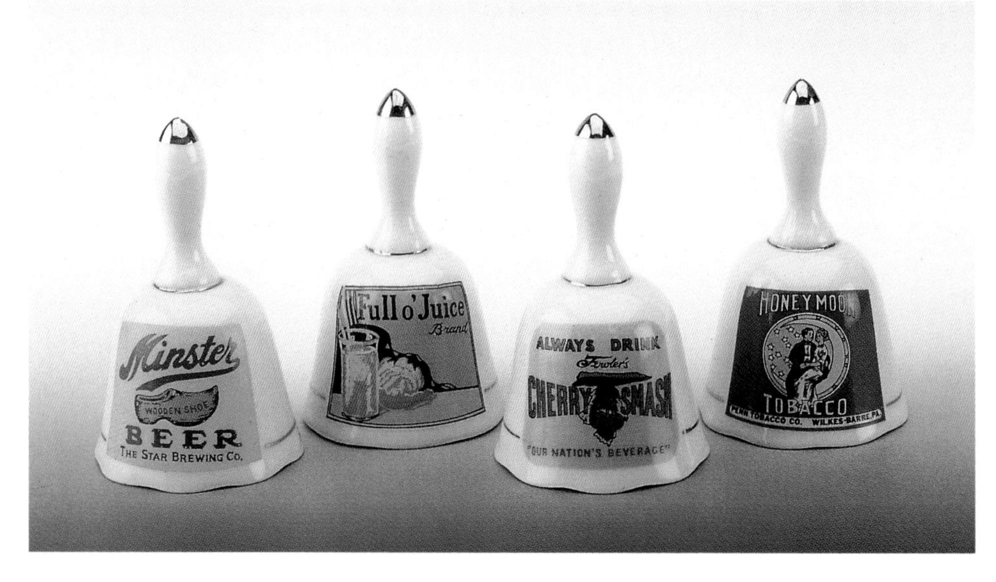

The first three of these small ceramic bells advertise your choice of beer, juice, or Fowler's "Cherry Smash." And perhaps those who partake would like to accompany their drink with a smoke from the Penn Tobacco Company, advertised by the fourth bell. 4" high. $3-5 each.

Advertising bell for Hoffman's at 34th and Broadway, probably a restaurant in New York. 4" high. $8-10.

At the Hilton in Chattanooga, Tennessee, guests can sleep in old train cars that have been converted into rooms and bring home playful bells advertising this unique hotel. $5-8.

Custard glass advertising bell. This pale lemon bell with a painted gold band around the bottom and base of the handle advertises "The Bal Tabarin" in New York City. The exact nature of this establishment remains a mystery, however. The bell's owner researched the name and found that Trows New York Co-partnership and Corporation Directory for 1911 listed a Bal Tabarin at the West 29th Street address shown on the bell, but doesn't say if it was a hotel, restaurant, etc. In the early 1930s there was a New York restaurant called Bal Tabarin, but it was located at 225 West 46th street, a different address from that shown on the bell. 6.25" high. $50-70.

An interesting bell from West Germany, probably advertising some type of beverage. The bell base appears to be made of hammered tin. 5.5" high. $5-8.

The scarecrow perched on top of this metal bell makes it an appropriate decoration for Halloween. 5" high. $30-35.

Seasonal and Holiday

Bells are a natural accompaniment to the festive decoration and joyful atmosphere that characterize so many of our popular holidays. While their sound and sight add a colorful note to any celebration, it is hardly surprising that the most popular holiday for bells is Christmas. The melodious ringing of bells has long been associated with both the religious and secular observance of Christmas; mighty church bells have heralded the arrival of Christmas since Medieval times and bells play a role in Christmas traditions all over the world. And when it comes to decorating the tree, there is no paucity of bell shaped tree ornaments to choose from!

Also perfect for Halloween is this ceramic ghost bell holding a lantern. The clapper is a simple silver chain. 4" high. $8-10.

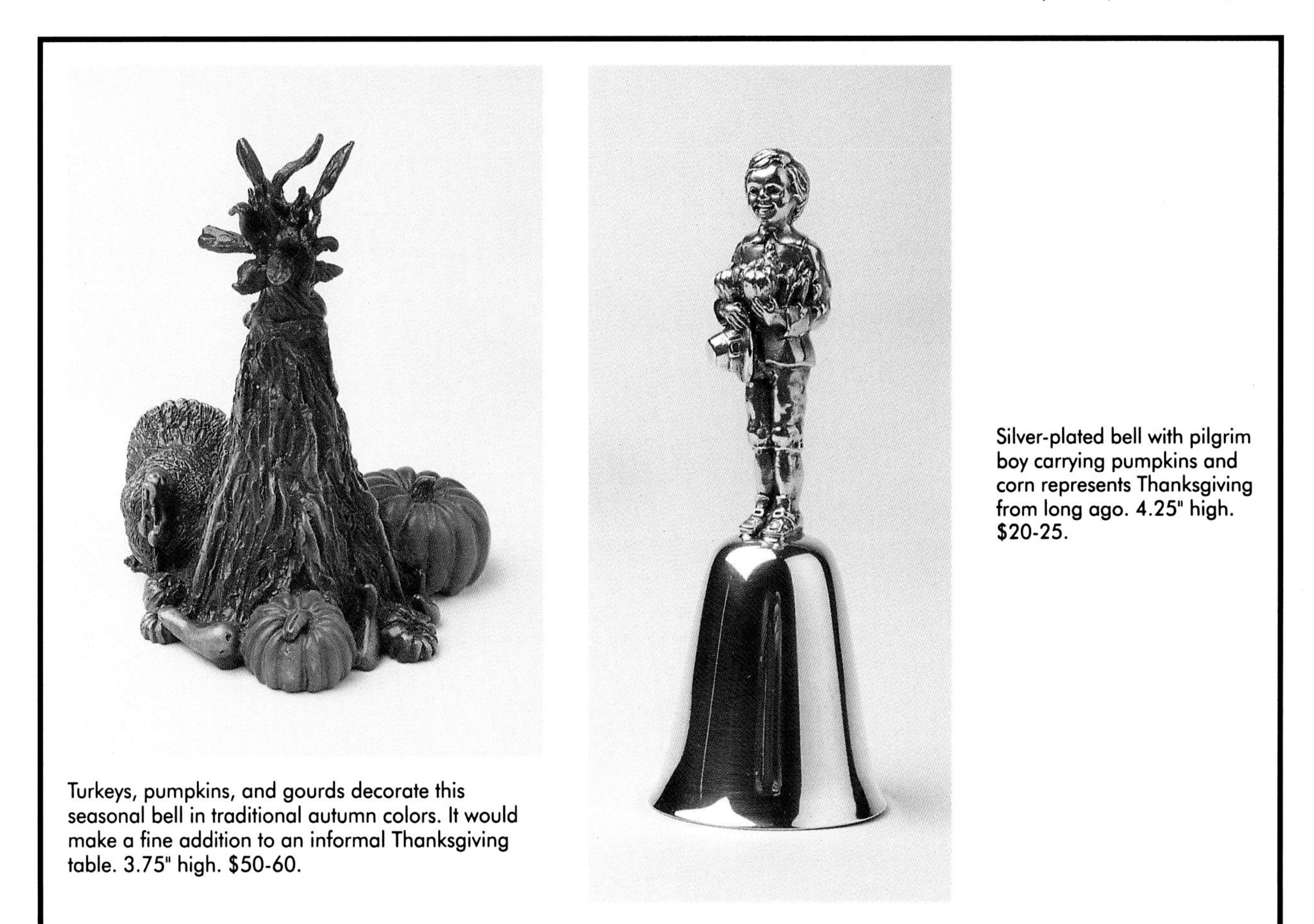

Turkeys, pumpkins, and gourds decorate this seasonal bell in traditional autumn colors. It would make a fine addition to an informal Thanksgiving table. 3.75" high. $50-60.

Silver-plated bell with pilgrim boy carrying pumpkins and corn represents Thanksgiving from long ago. 4.25" high. $20-25.

Silver-plated figural bells from a Reed and Barton series related to American holidays. These two show a pioneer mother holding an infant, representing Mother's Day, and the legendary Johnny Appleseed, representing Arbor Day. On the back of each bell the artist's name (Adolfo Procopio) is engraved. Pioneer mother: 5" high. Johnny Appleseed: 4.625" high. $20-25 each.

A third Reed and Barton holiday bell, this one representing St. Patrick's Day. 4" high. $20-25.

Pair of Christmas figurines of painted ceramic. 4" high. $5-8 each.

Hallmark series of Dickens caroler bells, made of fine porcelain. From left: Lord Chadwick, issued 1992, 4.75" high; Mrs. Beaumont, issued 1991, 4.5" high; Lady Daphne, issued 1993, 4.5" high; Mr. Ashbourne, issued 1990, 4.25" high. $30-45 each.

Pair of white ceramic "Mr. and Mrs." snowmen bells, made by Lefton. 4" high. $5-8 each.

Red, white, and gold ceramic Christmas bells. The figurine on the left carries a gold cross, the angel figurine on the right has wings on his back and a Christmas tree on his robe. Both 4.25" high. $5-8 each.

Lefton China figurines decorated for Christmas, c. 1940s. 4.25" high. $5-8 each.

Two additional sets of Lefton China angel figurines. These ceramic bells with wings on their backs carry gifts and holiday decorations. The "fur" around the base of their robes has been highlighted with gold trim. 4" high. $5-8 each.

Colorful series of porcelain Christmas bells by Bareuther, made in Germany. From left: 1974, 1976, 1984, 1973. The 1973 bell is a Limited First Edition, the other three are Limited Editions. All 6.25" high. $25-35 each.

Sarna Christmas bell, 1974 Limited First Edition. Decorated with the three wise men on one side, a nativity scene on the other. Marked "Christmas Greetings, 1974" on the outside and "No. 771 Bells of Sarna, India (C)" on the inside. The clapper is plug shaped brass, with "1974" and "Bells of Sarna" incised on the bottom and sides. 8.5" high. $100-150.

Sarna Christmas bell, 1975 Limited Second Edition. Six sided brass bell with large lettered wishes for "Christmas Cheer Through The Year, 1975." Inside markings: "Bells of Sarna, India (C), No. 1768, Limited Edition of 3000." Same clapper as previous bell, except "1975" is incised on bottom and sides. 6.25" high. $100-150.

Small porcelain bell with Christmas tree decoration and matching candlesticks. Bell: 2.5" high. Candlesticks: 6" high including candle. $20-25 for set.

Sarna Christmas bell, 1977 Limited Edition. This one features scenes of Bethlehem and shepherds on either side, an angel on the handle. Outside marked "Christmas Greetings, 1977," inside marked "Bells of Sarna, India, Limited Edition, No. 251." The largest of the three Sarna Christmas bells, this bell's tone is not as pleasing as the other two. 12.25" high. $100-150.

Swedish Christmas bell, brass. 4" high. $35-50.
The inscription translates to:

> When during Christmas they ring
> The urging sound of the bells,
> Oh! may they bring to my heart
> The tidings of salvation from God!

Another Swedish Christmas bell, this one purchased in Stockholm and picturing the Madonna and child. 4.5" high to top of round handle. $25-35.

Fenton "Christmas Eve" bell. Frosted white glass bell with hand painted scene of a man and woman walking past a church. 6.5" high. $45-60.

Christmas bell from Italy with raised gold dot decoration, signed by the artist. Inside reads: "Christmas 1973 Hand Etched and Painted in Italy, by Veneto Flair, 1443/2000." 5.5" high. $40-75.

A simple but engaging Christmas tree ornament. 4.25" high. $8-10.

Two musical Christmas tree ornaments, hand crafted and hand painted by the artisans of Anri, Italy. Wooden bells with bas relief designs of children on front. Bell on the left is from 1976 and plays "Adeste Fidelis;" bell on the right is from 1977 and plays "O Tannenbaum." $50-85 each.

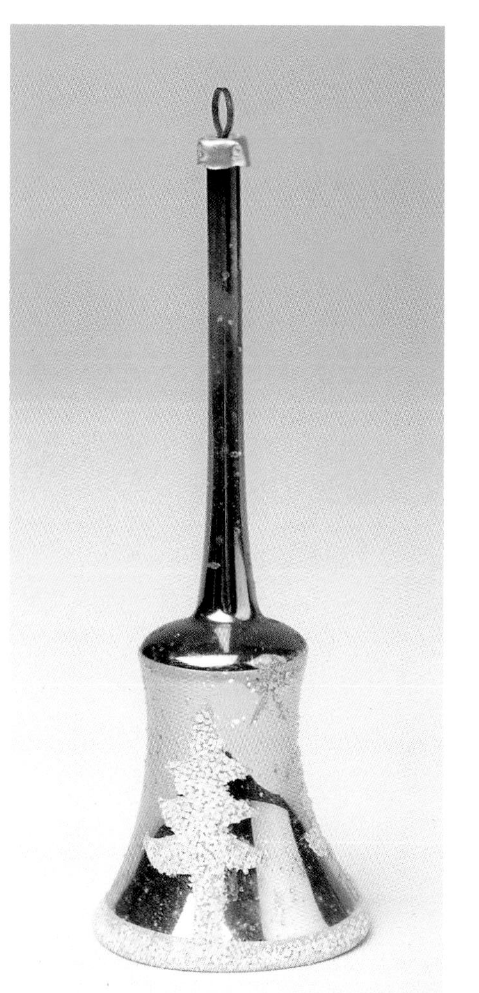

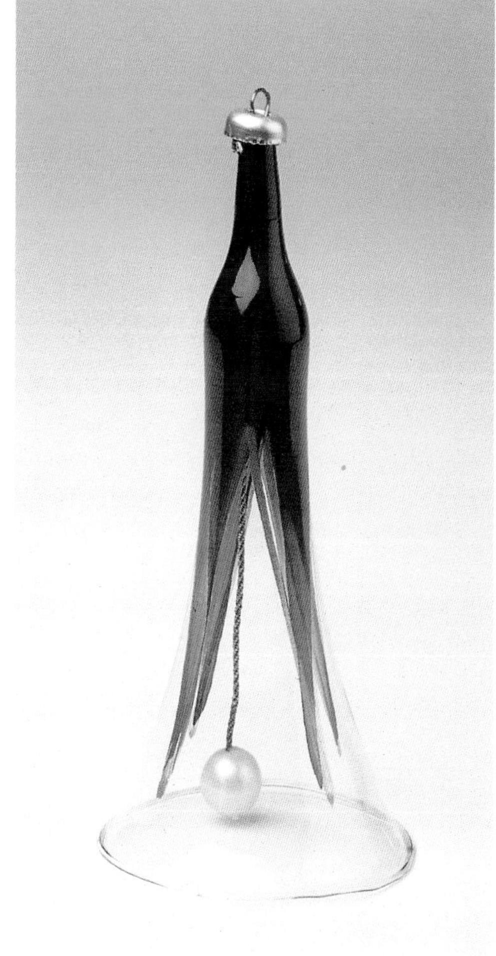

Right and far right: Glass ornaments for the tree. Green: 4.75" high. Red: 4.5" high. $2-5 each.

Conversation Pieces

Ranging from reincarnations of former objects to whiskey bottles disguising bells, this assortment of the unusual is sure to generate a second look.

A recycled oxygen tank was the inspiration for this bell, made by Vince Bowden of Bowden Bells in Minnesota. The tag attached to the bell describes Bowden as a machinist who found an artistic outlet in pottery. "'But I really hated pottery,' he says. 'Real men make things out of steel.' Thus Bowden Bells was born." Leather strap attached to clapper, beautiful tone. 13.75" high. $25-35.

Very heavy bell made out of a brass 90mm field gun Howitzer shell casing. Dated 1953 on the top; additional markings on the top include "WFD-2-36." 12" high, 4.5" diameter. Although purchased in Oregon, the origin of this bell is unknown. $30-40.

Can you identify the origin of this curious looking, gray metal bell? It was purchased in Minnesota and is made from the bowl of a milk separator. The handle and clapper are wood. 7" high, 5" diameter. $10-20.

This cowbell shaped "Alpine Bell" by Cesare 1970 doubles as a whiskey bottle. The photo of the bottle's underside confirms its status as a bell with a hollow interior and clapper. 9" high, 6.5" x 4.5" diameter. Value undetermined.

Eye-catching ceramic bell container for Lionstone Whiskey features a three dimensional depiction of the swallows of Capistrano. 9.25" high, 4.5" wide. Value undetermined.

Jewelry

As a decorative accent, bells are not limited to our homes. From the past right up to the present, individuals of both genders have used bells to embellish their clothing and their bodies. The jingle of small bells sewn onto clothing announces the comings and goings of the wearer and lends an atmosphere of merriment to even the gloomiest of days. Dancers in particular find bells a most befitting accessory to their costumes and a tuneful accompaniment to the music that defines their craft. As a fashion statement, bells have a universal allure:

> All around the world bells have been used on clothing. The natives of New Guinea made bells from shells to dangle from their clothing. In Northern Burma, the Naga women used to wear a short petticoat trimmed with bells, beads, and shells. On the west coast of Africa, it was the tradition for grown girls of the Benin to wear an apron consisting entirely of small brass bells. In parts of Asia, women wore stilt shoes trimmed with fringes of bells. Fourteenth century German gentleman jangled as they walked because of the bells trimming their jackets. And even in the twentieth century, bells were a part of the costumes of many teenagers in the hip scene. (Yolen 1974, 67)

Bell jewelry has a long history as well. Throughout the years, ornamental bells have bedecked toes, ankles, wrists, necks, and earlobes. We have already seen examples of the toe rings and ankle bracelets worn by Indian women in the chapter on Globetrotting. Many examples of bell jewelry from other parts of the world have also been chronicled, frequently belonging to individuals from royalty or affluence. Crotal shape bells were commonly used for bell jewelry, and in some cases these miniature crotals are among the smallest ones ever recorded. (Price 1983, 72)

Presented here are several selections from contemporary bell jewelry, worn by many bell lovers as conversation starters, day-brighteners, and visual proclamations of their affinity for bells. As jewelry designer Terry Mayer reminds us, the sweet sound of bells can certainly be "fashionable" in its own right:

> Wearing bells, as babies or brides or just for the fun of it, gives bells the chance to flirt in fashion. Whether an apple bell, pear bell, heart bell, puppy or teddy bear bell, star or flower bell, each in its fashion lets you make your own music. (Mayer 1992, S-13)

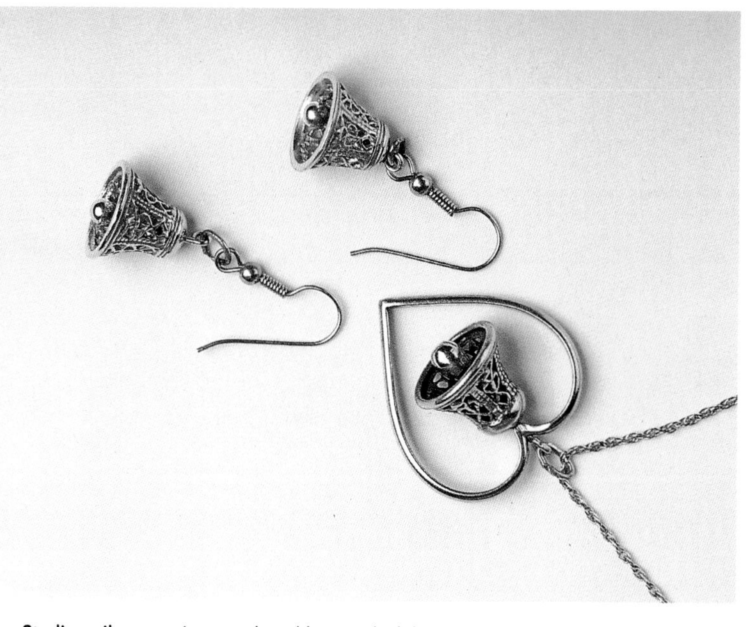

Sterling silver earrings and necklace with delicate filigree work. $50-75.

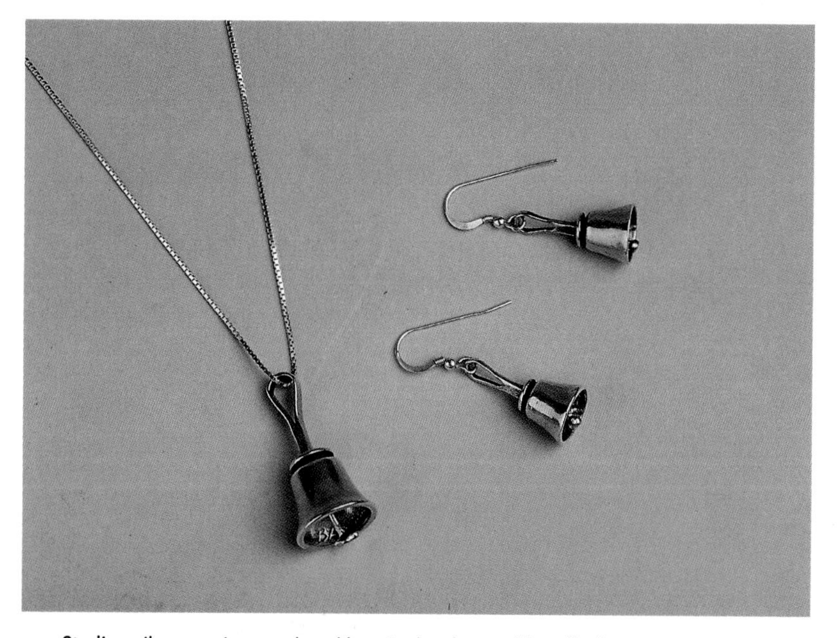

Sterling silver earrings and necklace in the shape of handbells, made by Bart Pass. $100-200.

Sterling silver earrings and necklace set, made by Gordon Barnett. $100-200.

Sterling silver cat design earrings and necklace, made by Terry Mayer. $100-200.

Sterling silver heart-shaped earrings and necklace set, made by Terry Mayer. $100-200.

Sterling silver earrings with bell and angel design. $30-45.

Floral cowbell set. Originally purchased as charms, made into earrings and necklace. $35-50.

Bell shaped earrings and necklace set from Korea. $40-50.

Liberty Bell earrings and necklace in gold. $50-65.

Earrings and necklace of Swarovski crystal bells with gold trim. Originally purchased as tiny bells, made into jewelry. $50-65.

Holiday set of Christmas tree bells with gold bead clappers. Gold filigree with colored stone. $65-75.

Gold earrings and necklace set with delicate rose design, purchased in Hawaii. $150-200.

Chapter Six

Special Mention Clappers

Not every bell has a clapper, but for those that do certainly its primary function is to produce sound by being swung against the bell's main body. In this utilitarian capacity, clappers tend to be of the "heard but not seen" variety, to update a common phrase. They are usually made of the same material as the bell itself, simple but pragmatic. In some cases the clapper is left in a crude, almost unfinished state.

Presented here are a few bells that take a different approach to clappers. For some, the clapper itself has the starring role. The Toriart crystal bell is nice enough, but without the engaging figurine clapper inside it's unlikely to rate a second glance. For others, the clapper's design bears a definite relationship to the subject of the bell and has been crafted with the same impeccable quality, often a tiny artistic gem unto itself. External clappers, such as that on the Pompeian crane bell, illustrate a style in which the clapper plays a more conspicuous part in the bell's overall design. And at least one clapper has a veritable story to tell, cleverly illustrating the results of a hungry dragon's nighttime snack!

Above and right: The clappers on this cast brass bell are a set of finely shaped legs and feet. Such leg-shaped clappers are generally indicative of figurines that are genuinely old. 3.5" high. $100-125.

Here is a contemporary version of legs serving as clappers. Set of four bell ornaments featuring characters from *The Wizard of Oz*. The unglazed porcelain figurines depict the Cowardly Lion, the Scarecrow, Dorothy with Toto, and the Tin Woodman. 3.5"-4" high. $35-40 for set.

Toriart "Wake Up Kiss" bell. Bell is made of clear lead crystal; suspended inside from a glued plate by a tiny hook is a Sarah Kay Anri figurine which acts as the clapper. 7.25" high. $20-25.

Two glass novelty bells with unusual clappers. The bell with the red plastic tractor forming its clapper is from Lancaster, Pennsylvania and the bell with the gold colored airplane clapper is from Oshkosh, Wisconsin. Tractor bell: 6.75" high; airplane bell: 4.5" high. $10-20 each.

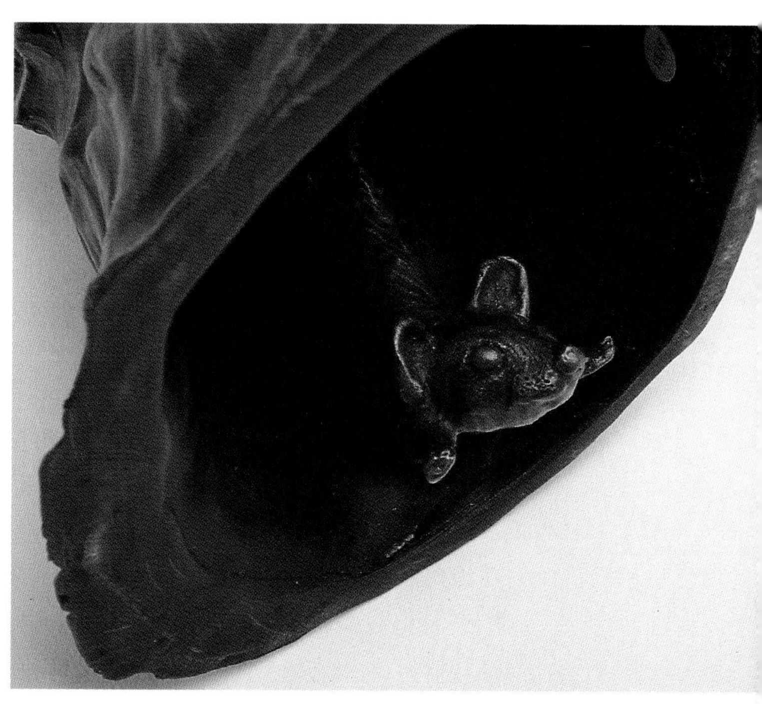

Great Horned Owl bell, by John McCombie. The dangling mouse clapper hanging from an S hook by its tail adds a bit of humor to the otherwise majestic owl with its sharp beak. The bell is initialed "JM 95" and also marked inside "#45 of 50 cast." 5.25" high to the ear tuft. $200-250.

The clapper of this Great Blue Heron bell, also by John McCombie, is a realistically detailed fish. See page 16 for another view of this impressive bell.

Pompeian crane bell with external clapper. This bell is a copy of one dug from the volcanic ruins after Mt. Vesuvius erupted in the first century AD. Atop the bell-shaped base stands a crane resting on one leg. The crane's other leg holds a chain with a bird on the end, forming the external clapper. Overall height: approximately 8". $25-40.

Top right and right: Ram's head bell, by Carl Wagner, with clapper in the shape of a hoof. 5.5" high. $200-300.

Carl Wagner's elephant bell (shown on page 143) also features a clapper shaped like the animal's leg and foot.

William J. Gerard, of the New York based Pewter Treasures, made this tiny bell and titled it "The Contented Dragon who had a Wonderful Knight." Turning the bell over reveals a horizontal flat knight clapper "swallowed" by the dragon. 1.25" high. $25-35.

These two bells are the work of artist Lowell Davis, a native of Missouri who portrays rural American farm life in his work. Both are "cold-cast" or sculptured porcelain, hand painted and signed. Their charm is enhanced by the realistic clappers found underneath. Left: Blossom the Cow, from the R.F.D. America Collection by Schmid, simulates a cowbell in shape and style. Clapper is a "metal" cow's leg, ending in a hoof. 4.5" high. Right: Old Blue and Lead Hounds, also from the R.F.D. America Collection. Clapper is a molded dog leg and paw. 3.5" high. $75-125 each.

Clappers from the Lowell Davis bells above.

Chapter Seven

A Bell Plus More—Multi-purpose Bells

Also a music box, this ceramic bell with Christmas doves and holly decoration plays the appropriate tune "Silver Bells." Made by the Ebeling & Reuss Company, 1985. 7" high. $20-25.

All of the items presented here are examples of bells coupled with some other object, the resulting merger both decorative and functional. While the bells shown are primarily of the open mouth variety, the combination lamp/calling card tray and the cast iron ashtray both feature mechanical style bells.

The two pair of bell candlesticks illustrate a common form of multi-purpose bell. In the shorter pair the bell actually serves as the base of the candlestick, with the candleholders acting as the bell "handles." The taller pair incorporates a small bell halfway up the stem, a more traditional saucer forming the candlestick base. A similar version of the latter kind was used in old English taverns: the candle helped guests find the way to their rooms, once there the bell functioned as a call bell to summon assistance. (Springer 1972, 208)

A particularly elegant type of multi-purpose bell is that known as a wedding cup, also referred to as a marriage cup. It consists of two bell shaped cups—a larger one formed by the skirt of a rather stylish woman and a smaller one held aloft by the same woman's arms. At weddings, the bride and groom both drank from the cups at the same time, the groom from the larger and the bride from the smaller. While the actual significance of this tradition is somewhat uncertain, the great variety of wedding cup sizes, styles, and materials attests to its apparent popularity.

Finally, the bell shaped vesta holder is an example of what Lois Springer terms "inanimates," described as "quiet objects having the shape and appearance of a bell but not functioning as one—that is, not ringing." Inanimates are also found in the guise of paperweights, banks, inkwells, biscuit tins, and a host of similar articles. (Springer 1972, 224)

Pair of tall bell candlesticks, made of brass. The bell is rung by pulling a chain attached to the rod which holds the bell. Overall height: 13". Bell: 1.5" high. $250-350 for pair.

Left: The bell hanging from a branch is part of the "appeal" of this ceramic decoration. A little girl sits next to a tree trunk, which also serves as a vase. Overall height: 6.25". Bell: 1.25" high. $10-15.

Pair of bell candlesticks with ornate foliage design. 3.75" high. $8-10 each.

Two gold-plated demitasse spoons with miniature bells on their handles. On the back of the handles is printed "Unique." 5" long. $15-20 for both.

Gold-plated sterling silver wedding cup bell. 6.25" high. $250-275.

This grouping of four wedding cup bells, with heights ranging from 8" to 3", illustrates the diversity of sizes in which these bells can be found. From left: Silver plate with gold wash in cup and inside skirt, $75-100; pewter, $65-80; silver with gold wash, $40-65; silver with gold wash, $50-150.

Combination lamp, calling card tray, and metal frog bell. The metal work is painted and the bell is triggered by a lily pad sticking out from the side of the frog's mouth.. The tray swivels and the lamp base simulates a tree limb. Overall height, 14". Frog: 7" long. Value undetermined.

Buoy bell paperweight with eagle on top. The rounded rubber bottom allows the bell to swing and strike. 5" high. $8-10.

Elgin clock with small Liberty Bell on top. The Declaration of Independence and portraits of John Hancock, George Washington, and Thomas Jefferson all decorate the clock face. Part of a commemorative series. $25-35.

Ashtray with twist bell. The cast iron base has two large receptacles for ashes and two smaller receptacles, possibly used for holding matches. The brass bell on top is rung by twisting the knob; the ring is quite loud! Around the base of the knob are the letters "R & E MFG CO NEW BRITAIN, CT. U.S.A." Cast in raised letters on the inside of the base is "RUSSEL & ERWIN MFG. Co. NEW BRITAIN CONN. U.S.A. PAT'D AUG.1.93." Also cast in incised letters under one of the ashtray arms is "R[D] No. 269895". Base: 7.5" long, 7" wide. Bell: 4" diameter. $70-90.

Above and right: Antique bell shaped wooden vesta (match) holder, covered with panels of tartan design. The panels alternate between Stuart and Murray tartans. This kind of wooden ware is known as "treen," and this item has quadruple collectible value (bell shapes, match holders, tartanware, treen). 4.25" high. $35-50.

Appendix

The American Bell Association

Those new to bells and those with an existing collection who want to expand their knowledge and understanding of bells will both find a valuable resource in the American Bell Association (ABA). This non-profit organization was founded in 1940 and has over forty-five regional, state, and international chapters that meet on a regular basis. The group holds an annual convention every June and publishes *The Bell Tower*, a bimonthly magazine featuring articles on all kinds of bells, collecting tips, pricing information, and chapter news. Some ABA members are particularly knowledgeable about certain kinds of bells—large bells, glass bells, or sleigh bells, for example—and are willing to share their experience and expertise through articles submitted to *The Bell Tower* or individual discussion with other members. The convention, held in a different location each year, affords bell collectors the opportunity to attend informational programs and develop lasting friendships with other bell fanciers. A highlight of each convention is the annual bell auction.

For further information, write to:

ABA
P.O. Box 19443
Indianapolis, IN 46219

Bibliography and Recommended Reading

Anthony, Dorothy Malone. *Bell Tidings*. N.p., n.d.
——. *World of Bells No. 5*. N.p., n.d.
——. *World of Collectible Bells*. N.p., n.d.
——. *The Lure of Bells*. N.p., 1989.
——. *More Bell Lore*. N.p., 1993.
——. *Bells Now and Long Ago*. N.p., 1995.
——. *Legendary Bells*. N.p., 1997.
Boland, Charles Michael. *Ring in the Jubilee: The Epic of America's Liberty Bell*. Riverside, Connecticut: The Chatham Press, Inc., 1973.
Clemens, Terri. *American Family Farm Antiques*. Radnor, Pennsylvania: Wallace-Homestead Book Company, 1994.
Collins, Louise. "The Way We Were...An ABA Retrospect." *The Bell Tower Supplement* 53, no. 3 (May-June 1995): S4-5.
Cook, Peter. *The Antique Buyer's Handbook*. Hong Kong: McLaren Publishing Hong Kong Limited, 1988.
Engels, Gerhard, and Susanne Sanderson-Engels. "The Bell Tolls: Foundry Technology in the History of Culture." *Foundry Management and Technology* 124, no. 4 (April 1996): 46-52.
Gaston, Mary Frank. *American Belleek*. Paducah, Kentucky: Collector Books, 1984.
Gelman, Sidney. "The Judaica Museum — Bronx, New York." *The Bell Tower* 53, no. 6 (November-December 1995): 22-24.
Glassco, Marjorie. "A Fredericksburg Belle." *The Bell Tower Supplement* 52, no. 1 (January-February 1995): S18-19.
——. *An Introduction to Bell Collecting*. The American Bell Association International, Inc., 1992.
Glassco, Marjorie, and Larry Glassco. "Curious Crotals, Part I." *The Bell Tower Supplement* 50, no. 6 (November-December 1992): S11-24.
——. "Curious Crotals, Part II." *The Bell Tower Supplement* 51, no. 1 (January-February 1993): S15-32.
Goeppinger, Neal. "Large Bell Basics — Part I." *The Bell Tower* 54, no. 2 (March-April 1996): 25-26.
——. "Large Bell Basics — Part II." *The Bell Tower* 54, no. 3 (May-June 1996): 15-17.
——. "Large Bell Values." *The Bell Tower Supplement* 49, no. 6 (November-December 1991): S2-4.
Hainworth, Henry. *A Collector's Dictionary*. London: Routledge & Kegan Paul Ltd, 1981.
Hammond, Lenore, and Curtis Hammond. "Collectible Glass and Porcelain Bells." *The Bell Tower* 46, no. 10 (October 1988).
——."Enameled Metal Bells." *The Bell Tower* 34, no. 11-12 (November-December 1976).
Hume, Ivor Noël. *A Guide to Artifacts of Colonial America*. New York: Alfred A. Knopf, 1985.
Kleven, Blanche, and Stanley Kleven. "Some New Thoughts on Collecting Figure Bells." *The Bell Tower* 51, no. 3 (May-June 1993): 20-41.
——. "The Source and Manufacture of Our Beautiful Bells." *The Bell Tower Supplement* 47, no. 3 (May-June 1989): B-T.
Mayer, Terry. "Even Fashion Rings a Bell." *The Bell Tower Supplement* 50, no. 2 (March-April 1992): S10-13.
McMillan, Gina. "Blooming Bells — Part I ." *The Bell Tower Supplement* 53, no. 4 (July-August 1995): S2-9.
Michael, George. *Basic Book of Antiques & Collectibles*. Radnor, Pennsylvania: Wallace-Homestead Book Company, 1992.
Price, Percival. *Bells and Man*. New York: Oxford University Press, 1983.
"Profile of a Bell Maker: Gerry Ballantyne, Bell Maker Extraordinaire." *The Bell Tower* 55, no. 5, (September-October 1997): 10-11.
Schick, R.D. "S.S.Sarna: A Mini-Biography." Special Report for Heart of America Chapter of ABA, Meeting at Auburn, Kansas, Sept. 20, 1981.
Schroeder, Don. "Identifying Characteristics of Sleigh Bells." *The Bell Tower* 44, no. 2 (February 1986).
Seki, Yasuo. "Story of a Bell Collector in Japan." *The Bell Tower* 53, no. 4 (July-August 1995): 12-13.
Shayt, David. "Elephant Bells." *The Bell Tower* 55, no. 5 (September-October 1997): 24.
Sieber, Mary, ed. *1997 Collector's Mart Magazine Price Guide to Limited Edition Collectibles*. Iola, Wisconsin: Krause Publications, 1996.
"Sleigh Bells—A Canadian Paper." *The Bell Tower* 54, no. 6 (November-December 1996): 26-27.
Springer, L. Elsinore. *The Collector's Book of Bells*. New York: Crown Publishers, Inc., 1972.
"Strange Odyssey of Mrs. Hall's U.S. Army Camel Bell." *The Bell Tower* 52, no. 3 (May-June 1994): 28-29.
The Special Bell Tower Committee, compilers. "1991 ABA Bell Calendar, The Second Edition: Description of Bells."
——. "1992 ABA Bell Calendar, The Third Edition: Description of Bells."
——. "1993 ABA Bell Calendar, The Fourth Edition: Description of Bells."
Thompkins, Jo. "Bells and Religion." *The Bell Tower*. 53, no. 2 (March-April, 1995): 23-40.
Trinidad, Al. "Ask a Question—Get an Answer." *The Bell Tower* 53. no. 2 (March-April 1995): 13.
Yolen, Jane. *Ring Out! A Book of Bells*. New York: The Seabury Press, 1974.

Index

About the Author

Donna Baker is a writer and editor from West Chester, Pennsylvania. She is a member of the American Bell Association International, Inc.

681.868848 B167

Baker, Donna S.
Collectible bells :
treasures of sight and sound

Central Business ADU CIRC

ibusw
Houston Public Library

5/09